DATE DUE

FORESTS,

FIRES,

by
Bob Gray

and
WILD THINGS

Front Cover:
Black Butte Lookout, built about 1936, was distroyed by the Columbus Day Storm on October 12, 1962. Later a new lookout was built (1963), and ten years later the new structure was moved to the Hogback Mountain on the Shasta Lake Ranger District by a Sikorsky sky crane helicopter, where it still sits.

ISBN: 978-087961-229-0

Art work and cover photos by Bob Gray

Books for a better world

Naturegraph Publishers, Inc.
PO Box 1047 ● 3543 Indian Creek Rd.
Happy Camp, CA 96039
(530) 493-5353
www.naturegraph.com
and Bob Gray
P.O. Box 121, McCloud, CA 96057

PAGE INDEX

PREFACE

"Ranger" as used in these tales includes everyone working for the Forest Service, whether he is a timber beast, engineer, mechanic, foreman, forester, lookout, biologist, landscape architect, construction worker, equipment operator, fire prevention technician, fuel manager, recreation assistant, snow ranger, or "somethin' else." A ranger is all of these, plus being an ecologist, economist, business manager, counselor, administrator, and housemother. He plans, he manages, guides, directs, leads, follows, and has a tremendous amount of responsibility to a multitude of people. His dealings with ranchers, loggers, hunters, fishermen, contractors, skiers, snow-mobilers, land owners, railroaders, hippies and spiritual groups, plus many other folks with various avocations, make the job a challenge, a pleasure, a frustration, and one of a great sense of accomplishment and unbounding interest.

The stories you are about to read are all true, and have happened to the writer, or to his associates over a period from 1942 to 1976. Names are real and it is hoped that no one is offended or embarrassed by the use of their names. The writings are dedicated to these good people who have been friends and co-workers over the years, and to my wife, Betty, and children, Douglas, Mary, David and Donald, who have shared many of my experiences, both good and bad. They also have shared many of my frustrations when I had to leave for a fire or a Forest Service mission in the midst of an anniversary, birthday dinner, graduation exercise, house guests or whatever may have been.

**Supervisor's Office
Shasta National Forest
1942**

MEETING DUTCH

A series of events led to my becoming a ranger in 1942. A punctured eardrum kept me out the the military service, and an ad in the Louisiana State University school newspaper for Forest Service workers in the western U.S. caught my eye. I had always had a notion that I would like to be a forest ranger, really not knowing what it was all about. My application resulted in an assignment to the Shasta National Forest, where I was to report to Mr. Davis, forest supervisor, at Mt. Shasta, California. An eventful hitchhiking trip from Louisiana ended in Mt. Shasta in late May 1942.

Mt. Shasta was a pretty town in a beautiful setting, making me think immediately that it should be my home. This later became a reality.

The forest service office was easy to find, and before I walked into the building, a man coming out the door asked where I was going. On telling him I had a job and was supposed to report to this office, he told me that I was to go to work for him in McCloud. This was after he made sure I could play softball, as there was to be a training school next week, and the highlight of the school was to be a softball tournament between all the ranger districts each afternoon.

This was a very fortunate happening for me, as Dutch was one of the greatest characters I have ever known; a rugged individual, not large, but powerfully built, weatherbeaten, smiling and with a perpetual pipe in his mouth. He was "The Ranger" at McCloud, a man of strength, wisdom, knowledge, and character, with an eighth grade education. A real old time ranger who had been and still was a terrific district ranger, but whose time was running out because of his limited formal schooling and the coming of age of the Forest Service.

By the way, we won the softball tournament that year, the trophy being a broad ax of ancient heritage, which the McCloud Ranger District kept for several years thereafter.

The supervisor's office never did find out why the kid from Louisiana never reported for work.

CACTUS & CLIMBING

"Cactus" was the McCloud District fire control assistant in 1942. He was rawboned, tough, and had real savvy in his job. His know-how in telephone communication, horsemanship, fire-fighting, road and trail maintenance, and equipment use were top quality. His strength, endurance, and self-assurance created a sense of confidence among all of us working for him.

Soon after arriving at McCloud, Cactus asked me if I could climb trees. Not quite realizing that he meant tall trees with no limbs for 40 feet and that I was to use tree spurs, I told him, "Sure, I can climb trees." What I had in mind was more like an oak tree or persimmon tree with lots of limbs.

"O.K.," he said, "let's get to work on the Grizzly Peak telephone line," which was a typical Forest Service line hanging loosely from tree trunks with staples, heavy wire, and split-ring insulators.

Well, I learned to climb trees again. Tree spurs feel like medieval torture irons after about the third tree, especially when they don't really fit, and when you're shaking from fear on the first day of climbing. The trees ran from six inch poles to six feet diameter trunks, with the insulators at about 25 feet above the ground, all of which required a different technique for climbing. At the end of the first day after climbing about 30 trees, your feet are numb, knees aching, and shins skinned, not to mention the aching arms from holding on and trying to get the job done.

The second and third days get better. As techniques improve, you get a better fit on the leg irons, and your muscles become accustomed to the new work. The confidence you acquire in the spurs also helps lessen the shakes when you're hanging to the tree by the safety belt, and using both hands to work with.

Like most training in the early 40's, it came on the job, with no one really telling you how to do anything.

Cactus could climb so easily and gracefully, that I determined to learn the job well, though I never did get as proficient as he.

TO THE LOOKOUT

In early June the fire season began, so Cactus and I started up the mountain to get me established in my summer job as lookout on Grizzly Peak. The plan was to take the Forest Service horse and 3 mules to the Hearst estate on the McCloud River and get one of the Hearst horses for me to ride the fourteen miles of trail to Grizzly. All was going well until Cal, the Hearst caretaker decided that their horses were not in good enough shape to make the long trip to the lookout. You guessed it, I walked. Now, fourteen miles doesn't sound like much, but the 3,500 foot elevation rise from the river to the lookout was quite a trip for a Louisiana swamp native, where the elevation of the ridges and hollows vary by about six feet. I guess the horse and pack mules were in worse shape than I, because my arrival at the lookout was a couple of hours earlier than Cactus and the stock. Some good directions by Cactus before leaving Hearst's kept me on the right trail. The instruction to take the forks that go uphill would have done the job. Yes, I was tired when I got there, but pretty proud of the accomplishment of covering the distance in about six hours. "You're a pretty tough kid, Louisiana," Cactus said, using a name he called me for a few years.

The lookout building was less than I expected, a ground level box ten feet square, with drooping shutters, sagging door, cracked windows, and a rusty gas cook stove. Peeling paint inside and out topped the first impression. Guess I had visualized a neat, white painted building on a low tower surrounded by trees, with a spring nearby. Oh well, live and learn.

We got the stock unloaded and fed, and all the supplies stowed away in the building before Cactus gave me a briefing on how to use the firefinder and report fires. I would have to learn the map on my own.

Cactus and the stock left early the next morning, after telling me the telephone line was broken somewhere between the lookout and Stouts Meadow. He suggested I try to fix it because it might be several days before he could get to it. In the meantime if I had a fire to report, I'd have to run three miles to the iron telephone at Stouts Meadow to report it.

Well, I worked on that phone line for a couple of days. Got it to work, too. With that job accomplished, Cactus never showed up again for six weeks!

FIXING THAT PHONE LINE

That phone line took off in an entirely different direction from where we came up a couple of days before, so it was all new country for me. What I lacked in knowledge of telephone communication systems had already been written in lots of telephone manuals, none of which I'd ever read. So armed with a pair of slipjoint pliers, I headed down the line, which followed the trail from Grizzly Peak to Stouts Meadow, where the iron phone was mounted on a tree, and also where the road ended. This was about three miles from the lookout.

Snowdrifts up to eight feet deep still remained near the lookout, and patchy snow continued far beyond Stouts Meadow, even though it was early June. Well, I'd hardly left the lookout when the phone line, a single strand of number nine galvanized wire abruptly disappeared into the snow bank and reappeared some 200 feet away. This has to be the cause of the telephone failure, I thought, as I began a tug-o-war with the wire, and if you think snow is a soft, light, fluffy stuff, it just ain't necessarily so. It was hard and heavy as concrete, very unresponsive. "I'll get back to this later, when the sun warms it up," I said to myself.

From the snowbank, the line lay on the ground, over rocks, through brush, and occasionally hung limply in an insulator on a scrubby, windblown pine tree. After about a mile of following the wire, it suddenly appeared very slack, and then a broken end curled into the air above the brush. If there's one broken end, the other has to be close by, I reckoned, and sure enough, there about fifty feet away lay the other end. Cactus had said there must be a break between the iron phone and the lookout,, and here it was. Now all I had to do was to tie the ends together, neither end of which would cooperate worth a darn. I tugged, jerked, pulled, and thought bad thoughts as I tried to tie the ends together. First I could only get the ends within about ten feet of each other, no matter how hard I tried. Finally, I walked down the line in both directions to clear brush and limbs from the wire in order to get some slack. Another couple feet were gained, but still there was about five feet of space between the ends. I looked the situation over and saw there was a long curve in the line, which I realized would give me the extra length if I straighted out the curve. This I began to do by pulling the wire through several insulators, and then making a beeline toward the other broken end. This almost did it. Now I only lacked six inches from getting the ends together, but try as hard as I could, no way would I close the gap. In desperation, an idea struck. By tying an end to my belt on one side, and the other end to the other side, I figured I could rotate my body and get more slack, kind of like using my body as a spool to roll the wire up. Sure enough, it worked. When I unwound, the ends were close enough to tie together, though the pliers didn't make a very neat job. The fact is, the knot looked much like a ball of spaghetti.

From this point on to Stouts Meadow there were lots of things wrong with that line, but no more breaks. When I cranked the iron phone and a voice answered at the ranger station, the surprise was apparent at both ends. Still we didn't know whether the phone would talk from Grizzly but I headed up the hill with high hopes.

Upon arrival at the lookout, I'd never been so tired in my life, but felt pretty good when a "Hello" came on the phone the first crank, but—something must still be wrong, because the voice kept saying hello, unable to hear my voice. Lee, the dispatcher, was the voice at the other end, and he kept ringing me, still not hearing my answer. What frustration—all that work and still no communication. After about an hour of staring at the contraption called a telephone, I made a discovery. There was a little button on the handle which must do something.

This time I squeezed the button when I talked, and the message went through! That phone was really a relic, in a brassbound wooden box, with a French style handset hanging on a nail. The handset was brass, and must have been a model of about 1880. So far as I know, no attempt was made to repair the line any further that summer. Those old Forest Service ground circuits were amazing in their efficiency and simplicity.

ANIMALS AT THE LOOKOUT

Rattlesnakes were plentiful. Of course, one or two of them are plenty. In the early summer they showed up at the most unexpected moments. The first one greeted me about 10 o'clock one morning as I stepped my bare feet out the front door while holding a pan of dishwater. Instinctively, or at least with little thought, I drenched him with the suds as he slithered away into the rocks. This one got away, but several others were disposed of with a long handled shovel before the summer ended. There were the remains of an older lookout building about 50 feet from the one I occupied, which proved to be the home of about ten baby rattlers when I tore it down for firewood and kindling. These were only about a foot long when I found them, but were really on the peck, kind of like a bunch of Bantam roosters.

Barefoot was a way of life at the lookout, but a wary eye checked the area before stepping out the door each time.

Quite often I spotted snakes while going to the spring about one-fourth mile away. No, I wasn't barefoot on these trips, which were almost daily, with a five gallon backpump to pack back up the hill. These trips kept me in good shape, as the climb from the spring was about 400 feet in elevation.

Grouse are somewhat less than the smartest creatures, with habits which can lead to an early extinction. On nearly every trip to the spring I would see a grouse or two, sometimes more. They would be running down the trail ahead of me, or perched on the same limb of the same fir tree most any time I passed. One day the thought occurred that one of those grouse might make a good meal, and since they sat on the limb like a bottle in the baseball throw at a fair, I'd try to knock one off. If I missed, they're so dumb they wouldn't fly away unless I hit the limb. Armed with three egg-sized round rocks I went grouse hunting. This time they weren't on the limb, or anywhere to be seen, so I started back up the hill. Then I saw one on the ground in front of me about twenty feet away. When I wound up to throw he decided to fly. That was his mistake, because he flew right into that rock, which would have missed him a mile. With the oldest grouse, and a pretty bad cook combination, I had the toughest, most unsavory meal I ever ate. No more grouse hunting after that!

Grizzly Peak is bear country, no longer inhabited by grizzlies, but plenty of California black bear of various shades, even one whose shaggy coat made him look like a tan and black pinto. Most of these were sighted about one-fourth to one-half mile away with binoculars, but one of them decided to check on what was going on at the mountain top. In late June, daylight lasts until about ten p.m. at 6,000 plus feet. I had gone to bed, still lying there awake when I heard a noise at the door. Glancing over my shoulder, I nearly had a heart attack as a huge bear stood up looking into the open front door through the screen. He probably was as startled as I, as he turned away to amble away down the trail. Had he been accustomed to humans as those bears in national parks, he could have been a problem, but real wild bears are quite fearful of men and he never came back.

Evidently all animals have curiosity about man. Even a big buck appeared one night, and I saw his large antlers scraping the window and side of the building just beside my bed. The building had no catwalk as most lookouts do, and the buck was walking on the ground which was a narrow ledge between the building and the sheer dropoff into Devil's Canyon on the east side. My movement and the flashlight beam frightened him away in a hurry. Does and fawns came by the saltlick quite regularly, but only one other buck came into view of the lookout during daylight hours all summer long.

Chipmunks, golden-mantled ground squirrels, and digger squirrels abounded in the area. One gray squirrel became friendly, but never really tame. Much of my food supply went toward these little rodents, which added company to an otherwise lonely job.

My favorites were the golden-mantles, which were the easiest to tame and became noisy and belligerent when they felt I was neglecting them. After a short time there were about six of these which I could hold in my hands, put in my pocket and let climb all over my head and shoulders with the reward of a stale biscuit or piece of bread. The chipmunks, little sharp-nosed varmints, were never real tame, but would eat out of my hand, even search my pockets, but never let me pick them up. The digger squirrels (also called ground squirrels) became quite tame, and would let me stroke their head and back while feeding them, but never liked to be picked up.

The gray squirrel was a real challenge. For weeks he would watch me from the rocks about 15 feet away, and would eat what I left out for him. After several weeks he would come to me by following a trail of crumbs, and eventually eat from my hand, but always holding my fingers with his front feet. Sometimes I could stroke his head with the back of my hand, but not the palm. He seemed to know that I couldn't grab him so long as I was using the back side.

One morning I heard a noise in the cupboard and found a chipmunk in the flour container. I had left the lid off, and it was funny to see the furry streak of white with a dust cloud of fine flour as he took off for the front door. Needless to say, I only used that flour to cook for the little ones after that.

During late summer the squirrels and chipmunks were not so tame, and I feel they were trying to become less dependent as winter and hibernation season approached.

BACKPACKING SUPPLIES

We brought so much food to the lookout on the first trip that six weeks went by before I had to have any more. By this time, fresh fruit, fresh meat, butter and eggs seemed like a dream come true. Cactus had called and said he had work to do at Stouts Meadow, and if I wanted to come down with my packsack I could have all these good things. It would still be a week or two before he packed in bottled gas and other supplies, so I told him I'd be there. The trip down the trail was like a vacation after the six weeks at the lookout. When I reached the meadow, Cactus was there with the fresh food, which was more than I could get into my packsack, but by coincidence, he had a "light" sheet metal stove which he suggested we put the groceries in, and tie the whole thing onto my packsack and back. I'd been doing without a heating stove up until this time, so agreed to pack it on my back. (I was about as sharp as a grouse in those days.) That load probably wasn't more than sixty pounds, but the balance and distribution of weight was pretty bad. A mule would have balked at the start! Not being a mule, I started up the hill. When I'd start to get tired I would back up against a tree and lean into it, or sit on the ground with the pack on a log or rock, with sweat pouring off my brow. Several times I thought I'd unload the stove and just pack what I could the rest of the way, but each time, lack of common sense prevailed. By the time the halfway point was reached, all my dozen oranges were gone, that juice was the best stuff I'd ever tasted. About 400 yards from the lookout, the trail made a big curve to gain the last elevation, but by climbing up the rocks, I figured to shorten the trip by about half. The planning wasn't bad until I came face to face with a small rattlesnake on a ledge about a foot from my face. At that distance he looked pretty big, but as most snakes do, he retreated into the rocks and let me by. That unbalanced load and needing both hands had left me in a bad position had the snake not had a place to go.

All's well that ends well, and the fresh fruit and supplies, plus a day to remember, made it a good venture.

Black Butte Lookout

A LOOKOUT'S LIFE

Accomodations and conveniences at lookouts vary from primitive to plush. Grizzly Peak in 1942 was definitely in the first category. Among the conveniences it didn't have was refrigeration. Until mid-July, the snowdrift nearby served well, except that it wasn't rodent and insect proof. Neither did it have water storage facilities, only the backpump and a few buckets to keep water in. Storage space was not adequate. The bed was a disaster, springs were broken, sagging, and rusty. the mattress (pad) was a compresssed slab of cotton in ticking, somewhat harder than the floor (which had a little spring to it). The cookstove was propane gas, with an oven that worked in spite of being totally rusted. Gas was packed up on mules in small cylinders.

The more accessible a lookout was, even in 1942, the better the facilities. Black Fox, some 12 miles from Grizzly, with road access had everything by comparison. For instance, Black Fox boasted an ice box, with snow in early summer and ice brought up regularly in late summer. Gas refrigerators came at a later date. A concrete tank held 500 gallons of water, which could be handpumped into a sink in the lookout cab. A garage and storage room was the bottom floor, a guest bedroom the second floor, and the top floor contained the lookout's quarters and work area. A short-wave radio and telephone made up the communication system. The firefinder, a device in the center of the cab has changed little over the years. A map on a 20 inch diameter steel plate, with a rotating sight or alidade, with the 360 degrees of a circle on the outer edge, describes the firefinder. The firefinder, map, binoculars, and daily log book are the main working tools of a lookout. Radio and telephone provide communication to other lookouts, mobile equipment, the ranger station and the dispatcher's office.

Procedures in reporting fires, relaying messages, ordering supplies and schedules of reporting weather, wind, cloud conditions and the like are outlined in policy and procedure handbooks which are given to each lookout.

Routine daily work can get downright monotonous. After the chores are completed, it's hard to stay alert hour after hour just looking for smoke, which may not show for days or even weeks. Recording weather forecasts, relaying a message occasionally, or washing the windows doesn't take up much of a sixteen hour day. Some lookouts read, sew, study, play the guitar, write, listen to the radio or even watch T.V. during the work day. All of this is O.K. so long as a systematic scan of the country is made at ten to fifteen minute intervals during the work hours. A few lookouts never stop looking, even while cooking and housecleaning.

Some lookout locations overlook areas of high fire occurrence. Bradley Lookout is one of these. Within its view are two towns, two settlements, an airport, 25 miles of railroad, powerlines, an interstate highway, many miles of rural roads, several resorts, a heavily used fishing stream, active construction projects, many isolated residences and several logging operations. All of these can and do cause forest fires. Where people are, fires occur. Consequently, it's easier for a lookout to keep alert at Bradley than, say, at Black Rock where towns, roads, transmission lines, railroads and all the man-made improvements are missing. The lookouts at these remote stations eagerly await a thunderstorm to liven up their activity, since they know that's probably the only chance they'll have of seeing and reporting a fire.

A thunderstorm at close range from a lookout is a beautiful and sometimes terrifying sight. All lookouts are protected from lightning by an arrangement of copper wires starting at the top of the roof, running down all gables and hips, along the edge of the eaves, down the corners of the building, into the ground at the foot of all four corners. The underground wires are buried for a distance of one hundred feet or more. Onto this maze of wires all the metal objects in the lookout are grounded. This includes the stoves, refrigerator, firefinder, bedsprings and any telephone line or plumbing in the building. In addition to these precautions, the lookout is instructed to disconnect the radio antenna, the telephone line and to stand on an insulated stool in the middle of the room while a storm is in progress at the lookout. Several lookouts have reported heavy jolts knocking them from the radio or telephone, but I know of no serious injuries to lookouts on the Shasta-Trinity Forest. On two occasions I've seen blue fire along the lightning protection wires and felt or heard a buzzing or humming sound from the wires during a storm.

Charlie, a lookout on Bradley took several time exposure photos of an approaching storm in 1975. One is a most striking picture of eight downstrikes over the Soda Creek, Squaw Valley Creek area. The strikes were ahead of the rain which followed in torrents. The cloud bases were illuminated as were the ridges and mountain tops in the distance, even though the picture was taken at 10:30 p.m.

When fires are in progress, a lookout's job is anything but monotonous. He or she feels a sense of pride or accomplishment in reporting the fire or fires, and a sense of responsibility to those fighting the fire. They report to the fire boss any change in conditions on the fire which will aid him in control of the fire. Spot fires out ahead of the main fire are reported. Wind shifts, smoke color change and whether the smoke is rising lazily or mushrooming is of importance to the man on the ground. When the fire is out, a lookout has usually played a big part in the success or failure of the suppression forces. A lookout quite often guides the troops into a fire by use of radio, mirror and knowledge of the terrain.

Some lookouts have many visitors, some few, depending on location and accessibility. This can vary from a half dozen to several hundred. Sometimes the person at the lookout makes a difference, especially when it's a cute young 18 year old girl on a pretty inaccessible peak who had all the boys on the fire crew volunteering to take her mail up most any time. I made a quick trip up once, too, but that was when the dog got "snake bit." More about that in another story.

Judy, the wife of lookout Charlie, wrote a song, "A Lookout's Life is Something Else," which covers the lookout's life in the mid-seventies. Here's what she wrote:

A Lookout's Life is Something Else

Hauling water up the stair, Putting up with dirty
hair and the radio that's always there.
   A lookout's life is something else.
Mixups on the grocery list. Leaky roof that's
never fixed, Who was that 10-8 I missed?
   A lookout's life is something else.

Washing windows windex fumes. Fourteen people crowd
the room; Everyone's 10-7 but you.
   A lookout's life is something else.
Swarms of questions all you hear, that never change
from year to year; Don't you get lonely way up here?
   A lookout's life is something else.

Watching thunderstorms at night, Watching for that
special strike, Knowing the next one might be right.
    A lookout's life is something else.
Looking 'till your eyeballs burst, trying not to think
the worst; That someone else might see it first.
    A lookout's life is something else.

Hauling water by the pail, clinging to the catwalk
rail; Can I survive the outhouse trail?
    A lookout's life is something else.
Wherever did my paycheck go? Should I call the
F.C.O.? Who will fight my cause below?
    A lookout's life is something else.

I finally turned in some smokes after my storm finally
broke, was that someone's idea of a joke?
    (I wonder why I even spoke.)
Sewer pond pump smoke ain't my style, nor incinerators
nor old leaf piles. But I'll turn 'em all in so Control can smile.
    A lookout's life is something else.

I'll not mention any names, We all know anyway—
just the same I wonder who thought up this smoke bomb game.
    (I wonder who will share the blame.)
    Next one I see from wooded knoll, I think I'll
call up Control, And have 'em send the bomb patrol,
    sit back and watch the tankers roll.

I don't know but I will bet, they're trying to pay
off the national debt by closing towers right and left,
    they haven't gotten mine yet.
Who but me could begin to sing, of the comfort that
my tower brings. I hope it will be here next spring.
    'Cause a lookout's life is something else.

GUARD SCHOOLS

My first was in 1942 at Camp Leaf, an old C.C.C. Camp, near Tennant on the Goosenest Ranger District of the Shasta. It was a real good school. We were taught to use "S" set radios by Tom, a radio technician who could make a cigar box talk. Come to think about it, the "S" set pretty well resembled a cigar box. The antenna was a 15 foot piece of wire with a rock or stick tied to the end to toss over the nearest tree limb. A toggle switch turned it on, and a variable dial was used to search out a transmitting frequency and a receiving frequency. The dial was usually well marked with pencil marks, because the frequencies never were the same two days in a row. A little luck and lots of perseverance were needed to operate one.

Cactus taught us how to use an ax and how to operate a Pacific pump, while Dutch and Henry gave instruction on fire line construction and fire behavior. Jack gave some history on the Forest Service, and the assistant supervisor of the Klamath talked on the effects of fire on the environment. There was lots more, but everybody's high points were the softball games after school. This was the first time the McCloud District won the broad ax trophy over the six other districts. Dutch had recruited well. The ball players were also pretty good fire fighters. McCloud kept the ax for the next three years, and it became their permanent possession.

Jack was the ranger on the old Redding District, a real good friend of Dutch. He was just as proud of his people as Dutch was of his, and there was lots of competitive spirit between them. So after McCloud won the softball trophy, Jack challenged Dutch to a tug-o-war between the districts, trying to gain some revenge for the softball loss. Well, the tug-o-war teams did a little conspiring and decided to make the tug-o-war contest a draw, and force Dutch and Jack into a boxing match. Neither of them ever knew that the draw was set-up, as they gamely put the gloves on. You know, that boxing match really was a draw. Jack had a bloody nose, and Dutch had taken some pretty hard punches before we stopped the battle.

Getting acquainted with your neighbors is one of the most

important reasons for guard school. Becoming familiar with those you're working with on fires later in the summer, makes a real difference in how well the team effort works in fire control.

LOOKOUT CHARACTERS

"A wide and sometimes wild assortment" best describes people who become lookouts. Background and education have little in common among the characters. I'd like to write briefly about some of these folks.

Al, who is still a lookout, was at Bear Mountain in 1942. Kind, gentle, softspoken, and sharp as can be, best describes Al. He was a bachelor, with a background in chemistry which he sometimes worked at during the off season. He is a self-taught meteorologist, having studied weather patterns for many seasons. In the early forties, the fire weather forecasters used his knowledge and observations regularly in forecasting thunderstorms and other weather phenomena. He can accurately locate fires and can describe the fire behavior and potential as well as any lookout I've ever known.

His communication with the little creatures on the mountain is unique. Once when I was at his lookout, he was outside with his family of chipmunks, including a mother and several young ones. A large hawk flew near, and instead of scurrying to the holes in the rocks, the whole family gathered around his feet, with the mother perched on the toe of his boot until the hawk had left.

Al is a talented photographer, with beautiful sunset pictures, some of which he used on Christmas cards each year. Since leaving Bear Mountain he has worked on several lookouts in Oregon and Idaho.

Leila, a retired pharmacist, became a lookout on Little Mt. Hoffman during World War II, when women first started being lookouts. She was a bachelor lady and no stranger to mountains and forests. Much of her life was spent in hiking and exploring the high country of the Sierras, usually with her friend Helen. She was a no nonsense type of person with a stern countenance, who took her job very seriously and never missed a fire. She must have never stopped looking. There were telephone lines between lookouts at that time, and if her smoke was closer to another lookout and they had failed to see it, she would ring them and say, "If you're not going to report that fire, I am." She really kept her fellow lookouts on their toes.

She wrote her lookout experiences into a very interesting story, but I don't know whether it was ever published.

At the end of World War II she again retired, so the soldiers could get a job "which she really didn't need."

Lenore, a high school English teacher, reported to work in 1945 with her electric iron, electric mixer, radio and all the conveniences of home. No, Black Fox still doesn't have electricity. It took more than that to discourage her, though, as she continued to be a lookout until 1974, with a year or two out for European and Asiatic tours. A real outgoing person, interested in everyone and everything, she totally lived her job as lookout and teacher. After retiring from teaching in the late 40's, she continued work with "her students" who visited and corresponded with her. The lookouts which were her home were always full of company coming and going—friends, family and students from all over the world. The American Field Service, a student foreign exchange program, was a part of her way of life. Everyone was royally welcomed at her lookout, from the fire crew bringing wood to her exotic visitors from Europe.

She was a good lookout, but even good lookouts make mistakes. One of hers was in reporting a smoke near Masonic Rock. Just a tiny wisp, she reported, so my crew and I started looking. There's no denser brush in the country than that around Masonic Rock, a lava outcropping northeast of Black Fox. When we reached the rock, no smoke was to be seen. "Oh, yes, it's still there, just north of you a little way," she would tell us on the radio. "No, you have gone too far, come back toward me and veer to your left," she said after we found a place to flash her with a signal mirror. This continued for a couple hours until our patience had worn thin and our pants thinner. And still we couldn't find the fire. It was a long way back to the fire truck, and an hour drive to the lookout, but I figured we'd better see what she was looking at, so we did. I still couldn't see anything after getting there, but you know what, I let that woman convince me that a hazy spot between two tall trees on the crest of the hill was smoke. And there we went again, charging through the brush for another couple of hours in tattered pants and bleeding shins, chasing a will-o-the-wisp. In later years there would have been an airplane or helicopter out there looking for it, but 1945 was still in the primitive era on the Shasta.

Lenore and I have remained friends for these 30 odd years since then as she worked with me until 1974, when her Sims lookout was closed permanently. She was a person who some thought demanded a lot in her job, but I know her as one who gave much more.

Then there was Pearl, my cousin-in-law, who was and is just as her name implies. She and her husband Jess were on Bear Mountain and she was later on Grizzly Peak and Bradley after Jess died of a heart attack. After a few seasons she became pretty much the den mother, or Dean, of the lookouts on the Shasta. All of the other lookouts went to her for guidance, condolences and advice whenever anything happened, and she usually was able to help them solve their problems. All this was done during the evening gab sessions on the F.S. radio, which was supposed to be limited to 30 minutes after nine p.m. To satisfy the talking needs of about 15 lookouts, though, it usually took longer. Pearl was always willing to talk as were most of the others.

THE FLAG & OTHER CHARACTERS

"Aunty" Pearl was the lookout at Bradley my first summer on the Sacramento Ranger District. From the first day she went up in 1960, she started asking for a new American flag. There was no doubt it was needed, as the one from last season was faded and shredded to ribbons as it flew in the brisk breeze. "I'll bring you one next week," I promised for about three weeks, forgetting to do so each time. As she became a little disgusted with my failing memory I tried to console her with the statement, "Now, Pearl, that flag has 'character', like none other at any neighboring lookout." This did little to satisfy her, but led to much publicity about "the U.S. Flag with the most character."

Carl, the publisher of the *Dunsmuir News,* visited the lookout a day or two later, with camera and pencil handy. He mentioned to Pearl that it seems the Forest Service could furnish a better flag than that rag on the pole. "Oh," quotes Pearl, "Bob Gray says that flag has character."

Next Thursday the *News* carried a story and picture of the "Flag with Character."

With a gentle prod from the district ranger, I got Pearl a new flag, with all the stars and stripes, but absolutely no "character."

One of the first helicopter trips to open a lookout on the Shasta Forest took Pearl and her dog "Cubby" to Grizzly. This was the beginning of the end of the horse and mule pack strings common in the 30's and 40's.

Mary has a six year longevity record for Black Butte, the last pack-in lookout on the Shasta Forest. She still mans Girard, and has done so for the past 18 seasons, except for one season stay at Bradley, where all those visitors were just too much. "Give me Girard back," she said, "where a visitor or two per week is maximum." A very self reliant lady, who does much to maintain her lookout, painting, repairing and throwing rocks off the road.

Annie Oakley had nothing on Mary, who keeps the snake population at a low level with deadly aim from her rifle. There's always a new set of rattles on the firefinder. Not many lookouts have a mountain lion hide hanging on the wall, but Mary does. Let me tell you about it.

Just before dusk one late summer day, Mary called in on the radio saying, "Bob, I just shot a mountain lion that was after my dog. I think he's lying in the brush by the little house." My thoughts were, darn that woman, she doesn't know a bobcat from a mountain lion. These thoughts, fortunately were kept to myself, as my son Doug and I got a rifle and flashlight together to go finish off the cat or whatever. By the time we reached the lookout, a dark night was upon us. Sure enough, though, there was lots of blood on the ground and brush where the cat was supposed to be, with a distinct splattering of blood trailing into the brush. Doug and I took off down the trail, one with the light, the other with the rifle. After about a hundred yards, the blood trail suddenly ended under a tree. With the sudden fear that I was about to be pounced upon by some unknown creature, I flashed the light up into the limbs above me, greatly relieved to find no bloody animal staring at me. Another thirty feet through the brush we heard a rustling sound and our light picked up a wounded, but far from dead mountain lion. His ears laid back and tail twitching, but his lower jaw broken from Mary's shot. We hastily put a bullet through his skull and dragged the 70 pound lion back to the lookout. Until this time in 1962, I'd never seen a wild live mountain lion, though Mary had seen another one at the lookout prior to this time.

The hide is beautiful, and when I skinned it for Mary, the meat also was a beautiful color, though, I'm sorry to say, I threw it away. People told me later that it is very good to eat and I'd like to have tried it. People who have eaten porcupine and 'possum shouldn't be squeamish about eating panther.

Tiny was a perpetual, eternal and everlasting story. Something was happening at her lookout all the time. The first night after she was taken to her first lookout, a violent thunderstorm pounded the McCloud and Shasta Lake districts. I was dispatching in the Supervisor's Office at the time, and lookouts were reporting fires and suspected fires in rapid succession. That is, all but Tiny. During a lull, Merv, the forest dispatcher, asked her if there were any strikes in her area, to which she replied, "Yes, but I don't see any fires, just a lot of flickering little lights, like fireflies, all around me." After a dead silence for some moments, Merv said, "Tiny, why don't you take some bearings on those fireflies?" She did, and was on her way to being an excellent lookout for a number of years thereafter.

When I moved to Fall River Mills, Tiny became my property, or vice versa. Anyway, she worked for me, or was it the other way around? She never ordered anything but exactly what she wanted. A ball point pen, for instance, had to have a red outside. No other color would do, even though the ink was the right color. Back it would go to the store. I usually ran these little errands after work on returning from her lookout. Patience, Bob, you must have patience.

Good housekeeping was just the name of a magazine to Tiny. Try as I may for two seasons, I never got the point across to her. In exasperation one day I told her, "Clean the windows, Tiny!" To this she said, "I don't have time to clean windows." "What do you mean, you don't have time? You're here 24 hours a day!" I could sense she was getting a little upset with my demand, as she told me, "Bob, except for the time I'm cooking meals and eating, I never stop looking from morning until night, and that's the important thing. Washing windows distracts me from looking." She kept looking as I washed the windows.

At the end of one season, as we closed the lookout, Tiny told me one of her cats was missing. I asked how she could tell, as there had been nine or ten racing around the interior like motorcycles in the hippodrome when I got there. (Wildest bunch of felines I ever saw!) We looked and called to no avail. Nine out of ten ain't bad, I reckoned to myself, but Tiny was pretty sad. Well, we found the cat in a cupboard next spring. Dead, and flat and, fortunately, well preserved from the coldness of the long winter. Even Tiny grinned a little as I picked it up by the stiff tail like a furry frying pan. We hoped this was just the first of his nine lives.

One fall, the weatherman gave us a 48 hour advisory of a heavy wet storm coming in. Knowing Tiny's lifestyle, I figured to give her a day and a half to get her gear packed to come down, thinking this should be enough for anyone. But no, when I got there, Tiny was sitting with her feet over the edge of the catwalk, and arms on the top of the railing, crying. "Bob, I'm not ready, I just couldn't get packed." Determined not to get agitated, I told her to get into the pickup and I'd take care of things. The lookout was on a 30 foot steel tower. I backed the pickup to the foot of it, spread a canvas over the bed of the pickup and started tossing all unbreakables down into the bed using bedding and pillows for a cushion on the first layer. Then I used a rope and a rapidly contrived hook to lower the more fragile articles onto the padded bed. In record time the lookout was empty, and the pickup bed piled high. By tying the four corners of the canvas together, the load looked like a gigantic hobo pack. Even Tiny was amazed. "I never thought we'd make it!" she exclaimed, very pleased.

Truly an unforgetable character.

Francile knows everything going on, anywhere, any time. Besides having a C.B. radio of 23 bands, she has a Shasta net radio, California Division of Forestry net, Klamath net and other receivers. No matter how many are talking at once, she never misses a call. Herd Peak is important to two forests and the Division of Forestry, looking at parts of all three protection areas. The sagebrush and juniper country below her is one of the most flammable areas in Northern California. Fires there burn fast and furious like they do in Southern California.

Francile takes her job seriously and does as well as any lookout I've known. It's always a pleasure to visit her lookout. Pops, her husband, has recently retired from International Paper and now spends most of his time at the lookout where he helps keep Francile on her toes. They have a real homey atmosphere at Herd Peak.

There are lots of stories associated with Francile, and this is one of my favorites. Fortunately, only a small percentage of potential disasters ever materialize and this is one that turned out to be just an interesting tale.

The Bolam Fire was changing directions every few minutes, running first one way and then another. We had a bulldozer working a flank of the fire as close as the heat would permit the operator and swamper to work. They were both on the tractor when a 90 degree change of wind blew flame through the canopy, forcing both men to bail out of the rig. They lit running, and the dozer disappeared into the smoke, as it was still in gear. Their radio was still on the tractor, so they couldn't call in to anyone to tell us their problem.

I was line boss, and got a radio call from the air attack boss saying, "Bob, there's a dozer running loose just outside the fire, with no operator on it." After a moment of stunned silence I told them to direct me to the dozer. All kinds of thoughts raced through my mind. What happened to the operator and swamper? Will it steer itself into the fire? Will it plow through the railroad track about one-quarter mile away? Will it hit a car on the highway? What if it runs into a crew of men or into a tanker? How are we going to stop it? Well, I finally caught a glimpse of the tractor. The blade was at ground level, and it was moving erratically as it hit lava outcroppings causing it to change direction. It seemed to have a guiding hand which kept it out of the fire, and turned it before plowing through the railroad fill. I was still about 100 yards from the dozer when I saw a man approach the dozer from the rear and somehow climb over the moving tracks into the seat. By the time I got there he had the tractor stopped and was sitting in the seat breathing hard with sweat running down his face. Then I recognized him as a rancher who had a few cattle in the area of the fire. When I asked him to drive it back to the fire edge, some 200 yards away, he told me he didn't know how to operate it. "You knew how to stop it," I said. "Yeah, but I'm not going to run it!" he replied.

I drove it back to the fire edge and built some fireline until the regular operator and swamper showed up. Never have I been so happy to see anyone as those two men, not knowing whether they were dead or alive.

Francile got some of this action on a tape recorder before running out of tape. One of these days I'm going to listen to the tape and find out how long it took for all this to happen.

Grizzly Peak has had more than its proportionate share of character types. Don, in 1945, was one. Truck driving had been his work, but he was just out of the U.S. Navy when he went to Grizzly. I can't really remember how efficient his performance was but I do remember that he thought he looked like Kay Kaiser, the old-time band leader of the 30's and 40's. He did look like him, and not only that, he could sing in a very acceptable Irish tenor voice. Sometimes he'd sing to other lookouts over the phone or when he had visitors.

He was not much of a woodsman. He had gone to Grizzly via Stouts Meadow, a three mile hike, so he should have remembered the trail. Here's what happened to him when he was to meet Cactus at the snow survey cabin one summer day. From the lookout to the cabin should take less than an hour down the good trail. Cactus arrived at the designated time, but an hour, then two, went by. No Don. Now Cactus was the nearest thing to an Indian scout in the country. He found Don's tracks where they turned off the trail onto a freshly worked trail heading downhill toward the McCloud River, and this was only 200 yards from the Stouts Meadow cabin.

Cactus figured Don had a two hour lead, and darkness was approaching rapidly, so he organized a search for the next morning, reasoning that if Don hadn't turned back, he was heading down Star City Creek by now. The next morning two parties of four men each started the search, equipped with radios. One party left the Hearst Estate at daylight, heading up the McCloud and then up the Star City Creek drainage. The other left Stouts Meadow heading down the Star City Creek drainage. It was mid-afternoon when the party going downstream called the other party on the radio, saying, "He's just ahead of us, we see wet footprints on rocks along the edge of the stream." Five minutes later, the two search parties met, and there was Don, tired, hungry and happy to be found, right in the middle of the two search parties.

"Well, I always read that when you're lost, head downstream," Don said, when we asked him how he ever got to where we found him.

About 25 years later Don again worked on the Shasta Lake District as a lookout, and on Burney Mountain of the Lassen Forest.

A year or so after Don, there were George and "Pancho" on Grizzly. Now, I've had some pretty bad housekeepers as lookouts, but George was one who, when asked or told by Paul, "George, when are you going to clean up this messy place?" would answer, "Why, I just finished before you got here." Besides being a lookout, George claimed to be a mechanic, a blacksmith and man of many talents. His pipe tobacco was out of this world. My wife, after talking to George for long periods of time on the Forest Service phone line, commented that the aroma even traveled through the wires. She sometimes wondered why she wasn't on the payroll for entertaining lookouts by telephone for an hour or so each day. He reciprocated by giving her a special formula for making laundry soap from bacon grease. George really wasn't too bad a lookout, and "Pancho," his big wire-hair terrier was a dog worth remembering. Both were bachelors at the time and needed each other.

Grizzly was blessed with characters; a new one each year during the late forties and early fifties. Her name escapes me, but the "Lady with the Monkey" came when Mac was district ranger. She was something of a glamour gal; blonde, stylish, and her companion was "Bozo," a fastidious creature who never really accepted the one-holer as being up to his accustomed standards. He always peered carefully into the darkness before balancing himself to use the facility.

The fact was, he hated the situation from the time he was loaded onto the back of a horse to make the three mile trip to Grizzly. Paul, the packer, "Tony," the horse and "Job" the mule were all in an unhealthy mental state before reaching the lookout. Mac had gone along to teach "Lady" the art of being a lookout and his experience with "Bozo" on horseback was rather messy. Seems the fidgety horse made "Bozo" nervous, and he couldn't control himself until he reached the one-holer. This happened while "Bozo" was climbing Mac's neck. The new Grizzly lookout was being built during her stay on Grizzly, and the carpenter crew's lack of understanding with "Bozo" created some problems. Anyway, "Lady" and "Bozo" did not return to the lookout the following year, but it was a unique situation for a summer.

ON BECOMING A FIRECREW FOREMAN

In 1945 I became a full-time Forest Service employee. A season as a lookout was enough of that lonely life for a 22 year old, so after two years in construction in private industry, I applied again to the Forest Service. This time as a trail crew foreman and packer, because I have a way with animals. The job never materialized because Dutch fired the firecrew foreman while I was standing there and told me that I was going to be crew foreman at Bartle. This was a good break, except that I'd never been on a fire before and had never been a foreman, either. "Don't worry about that," Dutch said, "because Cactus or I will be there to take charge." In the meantime, Dutch gave me a two minute training session in fire behavior which went something like this.

"Remember, a dry stick burns hotter than a wet stick; fires burn uphill faster than downhill; and wind can play hell with any fire." With these words of wisdom I took over the foreman job at Bartle. What the crew and I lacked in knowledge, we made up in enthusiasm, as we had a busy and successful season.

The crew was made up of Bud, Rudy, Russ and me. All were 17 or 18 and I was 23, the only real reason for my being foreman, as they had all been on a fire or two in early spring before I came and, consequently, knew more about fighting fires than I did. Not wanting to bluff my way through, I told them that we were all learning together and I expected them to help all they could, and let me know if they thought I was fouling things up. Also, I'd be willing to listen and learn with them and from them. We got along well, and none of them resented being bossed by a Southerner who hadn't yet learned what he was doing. Being my first crew, makes their memory something special to me. I still keep track of them after 31 years. Bud is a millwright in Sweethome, Oregon; Rudy works for Southern Pacific Railroad; and Russ stayed with the Forest Service, where he is presently the district ranger in Brookings, Oregon.

Other years in the later forties, as crew foreman, I had a couple of dozen crewmen working for me. Some good, some not so good, old and young, some smart and some not, but all of them I remember well, as each had some quality which contributed to my life. Some of the stories to follow will include these men.

CREWMEN CHARACTERS

Bud was a resourceful and ingenious young man. Anything said or suggested would start his mind to working, with action immediately following. One day I commented on the terrible dust from the much traveled dirt road in front of the guard station. It was several hours later that a motorist stopped at the station asking what all those boulders were doing in the road about 100 yards from here. Yes, Bud had built an obstacle course for vehicles to pass through, at no more than five miles per hour. Much more effective than a speed limit sign would have been. Moving the rocks back to their original site kept him busy and out of trouble for the next two hours.

Rudy lacked the energy level of Bud. The fact is, he bordered on being lazy. There was nothing he liked better than an occasional siesta. Well, I made a point of giving Rudy jobs that kept him busy for several hours at a time.

When a fire was reported, most of the boys, including me, would really get our adrenalin worked up. Not Rudy. He could climb into the long seat in the back of the truck and sleep until we woke him up when we got to the fire. "I'm really rested and ready to work when we get there," he told me. And he really was, as no one worked harder on the fire.

Nothing pleased Rudy more than beating me to the easy chair on the front porch after dinner each evening.

Russ, who is now a ranger, probably was the most dedicated worker I ever had. Always plugging away at whatever we were doing. He and I had a checker tournament lasting two fire seasons. There were so many games, that we kept score by how many games ahead one of us was. We were so evenly matched that a ten game spread was the widest margin we ever had. If I recall correctly, he was two games up after two seasons. I've got to challenge him to a couple of games next summer when I plan to visit him for some deep sea fishing.

Jay was our cook, and one of the best. He just spent too much time at the beer joint about a quarter of a mile away.

About once a week he needed encouragement to get through the day, but the Forest Service was pretty tolerant with his problem because he was so good and so personable. I spent lots of time working with Jay. Very frustrating.

One morning about two o'clock, Jay woke me up, both of us bleary-eyed (me from sleep) and said, "Bob, I just shot a bear in the cookhouse; he's dead on the back steps." "Jay," I said, "go to bed, it's almost time for you to get up to cook breakfast." He must have sensed that I was not too happy, so he left. Well, I figured that with the shape Jay was in I'd better get up and fix breakfast, so I did at 5:30. Jay was already in the kitchen, firing up the old wood cookstove. He was really in good shape and good spirits when he said, "What are you going to do about that bear?" Sure enough, a 300 pound black bear was dead outside the back door, with a bullethole between the eyes. He never could tell me how he managed to kill that bear when he came in from the bar, but he insisted it was in the kitchen when he got in to go to bed, and he shot it with his rifle which he kept in his bedroom in the cookhouse. He must have shot it inside the building, because none of the crew, sleeping in a tent nearby, nor I, heard the shot!

WORK WAS FUN

Togetherness was the thing for fire crews in the 40's and 50's. Twenty-four hours a day and five to seven days per week was the rule. No, we didn't get paid for all that time, that was just the way things were. Sure, we could go to town on Saturday night to the dance, but all of us went and the firetruck was with us. Occasionally, if our hair got too long we were told to go to town and get a haircut. (That wouldn't happen today with the present hairstyles.) Getting along with each other under these conditions was probably the most important part of the job.

We had fun on the job, with little or no scheduling of work

hours, we always exceeded an eight hour day. Early June was spent logging roads and maintaining telephone lines to lookouts. The Stevens Pass Fire of the 30's had left about 60,000 acres of huge snags. Winter snows and winds caused many to fall across the roads and phone lines. Our job was to cut and remove these windfalls. Chainsaws were not common until the end of the 40's, so we became quite proficient with two man crosscut saws, or "misery-whips," which we also used on fires. We would try to make only two cuts on a log, then roll or drag the piece out by hand or with a cable and a rolling hitch, until it was off the side of the road. Sometimes we would log 40 miles of road, and sometimes two, depending on the number of logs. Everyone worked together in determining the best strategy in removing a big log. If we were very far from home at quitting time, sometimes we'd just camp out on the spot or the nearest place we could tie our portable phone to the line.

None of us were really good at tree climbing with spurs, but everyone tried, and before long became fairly good at it. It was a new experience, interesting for all of us, and each tried to outdo the others. Dutch and Cactus were pretty proud of our accomplishments and never failed to compliment us on a job well done.

While working the line to Medicine Lake, we camped one night on the lawn at Harris Springs and had a nice experience. Deer were so numerous in the area that we saw dozens of them daily, especially as they crowded around the salt lick behind the Harris Springs station. Deer behave differently in June than in September and October, especially when still bunched up as they come into their summer feeding range. Well, we had our sleeping bags spread out on a green grassy plot, which the deer wanted to feed on. At daylight we awakened surrounded by deer legs. They were grazing right up to our beds, and seemingly ignoring our presence. No bucks, but about 20 does and their offspring, which Oscar, the Harris Springs fireman, had salted and talked to over the years.

Cutting wood for the lookouts and our own station kept us in shape and plenty busy. Sometimes we'd haul the wood on top of our truck, which had stakesides around the slip-on tanker unit, and sometimes Dutch or Cactus would haul it from Bartle to the lookout . . . Do you suppose those guys were hauling it to their homes? On, heavens, no, there's a sign on the dashboard of all government pickups that says "Whoever uses this vehicle for personal gain is _____, and going to be in bad trouble."

After cutting many cords by hand, they got us a power saw made from an old Ford pickup and a heavy crosscut running from the transmission and drive shaft. All we had to do was move the rig up the log and watch it cut. A real improvement, just seeing a machine do all the work. The saw had four wheels and a tongue to tow with, but the tongue wouldn't fit our truck hitch, so we improvised by tying it on with some telephone wire, which worked well, except once. We got back to Bartle from woodcutting, and our saw wasn't tied behind. We went back about five miles to where we'd been cutting, but didn't see it anywhere. So, we turned around and looked more carefully on the way back down the road. Pretty soon we found it, about 50 feet off the road in dense, but soft brush. There was no damage, though it took us hours to get it back on the road. We rigged up twice as good a hitch from then on, by doubling the amount of telephone wire.

A banner fire year on the McCloud District was 1945, with a few man-caused ones, and dozens of lightning fires. Let me tell you about a typical lightning fire. First, it's usually pretty inaccessible, a long hike, uphill, and through wet brush, a combination which has you exhausted on arrival. Many times, too, they're hard to find, especially in flat country with tall timber. Fog patches, haze, and dense vegetation add to the difficulty of just finding them. I've passed within 100 feet of a fire and missed seeing it on more than one occasion.

Arriving at the fire bolsters everyone's spirits and makes the frustration of finding it just something to tuck away in your mind, because now the work really begins. You can't believe the amount of hard, dirty, and hazardous work involved in controlling and mopping up a quarter acre lightning fire. First the foreman decides what the crew is going to do, and tells the men how to go about it. Fortunately, most lightning fires are accompanied by enough rain or hail to prevent rapid spread of the fire, so the first thing is to scratch and dig a narrow line down to bare dirt or rock around the perimeter of the fire before the drying period starts. This involves cutting away brush, reproduction, logs and anything else in the planned line. All woody litter, and decaying vegetative matter must be removed to prevent fire from creeping across. While all this is going on, the foreman or crewboss is determining what to do with the huge snag burning from bottom to top of its 100 foot length. He's taking note of its size, stability, direction of lean, and where all the embers are flying, and at the same time telling the crew where it is not safe to work. Of all the jobs in the Forest Service or anywhere else, there's none more hazardous than fighting a small lightning fire with a big hairy snag on fire. There's no way to do this job really safely, but everyone involved must know exactly what is likely to happen and the move to make when it happens. Most frequently with a big snag, small limbs and chunks of bark start falling as the fire burns. Large limbs will later burn off and come down at about 50 miles per hour when they hit the ground with enough force to break a man's neck or back, even with a hard hat on. Weakened and burned areas may cause the whole top or even half of the snag to fall at once with a mighty "whoosh" and "thud", as a half ton of flaming wood hits the ground. The fallers, in the older days, with a two man crosscut and now a man with a chainsaw, might be working on the snag when this happens. A preplanned escape route has already been established, with a signal man to holler, or tug on a string tied to the fallers belt when something starts to fall. The signal man never takes his eye off the snag, and the faller doesn't dawdle, even a half second.

Now, we've got the snag down and a fireline complete, but the time-consuming job is still to come. Because of the remoteness and inaccessibility of the fire, we have no water to put the fire out. Different firemen have different techniques.

After I look the quarter acre over and decide what to do, we'll take all the flaming and smoldering solid pieces and pile them in one or more piles to intensify the heat, and burn up more rapidly. We might ignore one section of the fire because the fuel is light and should go out while we put the rest out. Now we will roll the larger logs over into a cool place, getting them out of their bed of coals, with the hot side up. A few shovels full of cool dirt rubbed into the top side of the logs will cool it rapidly, and the damp night air will finish putting it out. The duff and deep litter is another story. With shovels and pulaskis, the smoldering material is turned over and mixed with dirt which cools and separates the burning materials so that they go out.

Forty-eight hours later, the fire might be out. No smokes are showing and all ashes are cold. The crew waits another four hours to make sure no smokes appear, and heads for home. We're still not through, though, as the fire will be checked at least once more to guarantee it hasn't come back to life to escape when a wind comes up.

Larger fires are not mopped up to this extent, only the outside perimeter to a distance prescribed by the fire boss. They sometimes take weeks to burn themselves out.

For about three years just after World War II, deer hunting was fabulous on the McCloud District. At Bartle we would validate between 150 and 200 bucks each year. Day and night, hunters would stop by with their trophies. The area, then as now, was a three point or better zone, so we saw some beautiful large bucks. In those days we weighed the deer, measured the antler diameter and spread width, reported the physical condition of the deer, spotted on a map the location of the kill, and listened to many incredible hunter tales.

Most hunters overestimated the weight of their deer, especially when they had to pack or drag it a long distance. "Come on," they would say after we had weighted a deer in at 130 pounds, "he's got to be 175 pounds." Usually we could convince them our scales were accurate by weighing the hunter himself. Every time there was a lady in a hunting party, the tags indicated she had killed the deer, though the male members usually would have a mental lapse, saying something like "I shot him just as he cleared the brush," then pausing when he realized what he said, then adding, "Of course, the little woman was shooting, too, so we couldn't tell exactly who hit him."

A few forked horn bucks were brought in, the hunter not knowing he was in a three point district. We would have to confiscate the deer until we got hold of a game warden. One man brought in a 210 pound forked horn. He expected us to validate it, because of prominent eye guards on the antlers. I would have done so, except that Don, the game warden was right there. Any hunter would have shot that buck without counting points, because of the size of the deer and horns. The warden told the unhappy hunter to report to the judge at McCloud the next day, which he did. Don had obviously not told the judge, because he did not show up, thus, a compromise settlement was made. The hunter lost his deer but didn't get arrested. Don later told me that he would have given the deer to him, too, "If all you guys hadn't been around!"

Another large forked horn was brought in by a man, with the location of kill shown as Four Corners. There are two Four Corners locations, one in the three point area and the other in the forked horn area. This deer was similar to the other, with huge antlers and eye guards. I validated it, assuming that the Four Corners on the tag was the one in the forked horn district, believing that justice was served in this case.

It was in one of these years that a four point *doe* was killed near Medicine Lake. I didn't validate it, but the hunter stopped by Bartle and showed her to us. She was a medium doe, with delicate doe features, and a small rack on her head.

Several times hunters stopped by to tell me that an illegal buck or doe had been shot and was lying in a certain location, usually alongside a road. Our crew would go and pick up the deer, and use it at our camp. Probably a mistake had been made by the hunter reporting the deer, but we never pressed the issue. Better to salvage the meat than let it rot in the woods.

NOSE TO NOSE WITH A BUCK

Our saltlick was about 200 feet back of the Bartle Station and for a couple of weeks two big bucks had been coming to the lick at dusk. An idea occurred to me that I would like to see how close the deer would come to the salt while I was there.

The lick consisted of a salt block on a stump, with a single rail, pole fence, about 30 inches high. It would keep cattle out, and allow deer to either jump over or go under.

About a half hour before dark I sat down in the closure, with the stump to my back, and the salt block behind my head, facing the direction the deer always came from. The weather was still and warm, and a few flies and gnats were buzzing around; making a nuisance of themselves.

Right on schedule the two bucks showed up, with darkness rapidly approaching. They both got to the fence and stopped to check their surroundings, while I sat stonestill looking right at them. Deer have sensitive ears, but like most animals, don't see too well unless something is moving, so they hadn't seen me yet. Their noses, too, are sensitive, but the still air hadn't carried my scent to them. The larger buck, a four-pointer, jumped the fence and stopped about four feet from me, and the other one lowered his antlers, and crawled under the fence, but stayed behind the larger one. Old "four-point" stood like a statue, with his nose toward me, while "young buck" fidgeted in the background. My nose started itching, and the sweat was beginning to run down my forehead, but I did not dare move for fear of frightening the bucks away. That beautiful wild creature only four feet from me was breathtaking, and a little scary.

Suddenly that buck took two steps toward me, and was literally nose to nose with me. Wild thoughts raced through my mind. Is he going to ram a horn into my face, or cut me with his front feet, or what? It had reached a stalemate between us. I was scared, and I didn't know what he was thinking, surely he had something on his mind. We made our move together, both bucks whirled and cleared the fence as I rolled around the stump. You can't imagine how good it feels to scratch your nose when you haven't been able to for a while!

BARTLE

During the early history of the McCloud River Railroad, Bartle had been a much busier community than it was in 1945, or even 32 years later in 1977. In the early 1900's, it had even boasted a hotel for weary travelers. A book, *Pine Across the Mountain*, written by Robert M. Hanft is an excellent history of the McCloud River Railroad with many references to Bartle.

In 1945, the Bartle Forest Fire Station, a section crew foreman house across the tracks, and a redwood water tank for the steam locomotives was all that remained. The McIntosh Ranch and the Bartle Beer Joint were about one-half mile from the guard station. All these places were occupied only in the summertime, except for the section crew foreman's residence, in which Shorty, the section foreman, his friend Julia, and his, hers or their two children lived. I grew quite attached to the two kids, a boy about three and a girl about four. Real rag-tags they were, but cute as could be, and they really ate up the attention shown them by the fire crew. Bud was always indignant that Shorty and Julia didn't take proper care of their kids. "Lets 'em run around like wild animals," he growled.

The McCloud train goes right by the guard station, and in those days used to stop for water for the steam engines. Each morning one or two trains would go east with 20 to 40 empty flatcars, and return in the late afternoon loaded with logs. In 1945, the rumor was that McCloud had ten more years of logging before all the timber was cut. But 1977 has arrived, and they're still cutting 400,000 board feet daily, though the days of railroad logging are long gone.

The railroad was pretty loosely managed in those days. On several occasions, either I or one of the fire crew caught rides to McCloud in the engine cab, or they were very obliging to take mail or messages in for us. During hunting season, the cab of the engine, the caboose and several flatcars bristled with deer rifles. Some of the hunters would ride out to their favorite spot to hunt, and get back on the train as it came back in the evening. The train crew would shoot from the train and bring in lots of bucks.

At Bartle we had about eight to ten Forest Service employees. Three or four on the fire crew, three road crew members, and in 1945 we had a special crew of young men who really didn't seem to have a purpose. They would get up, eat breakfast and disappear until suppertime. They were supposed to be making some sort of special study in the forest, but they were very quiet and uncommunicative, spending most of their spare time reading the Bible. They were really nice young men. Being a nosy, curious person, I finally asked one of the men why they were the way they were. Well, in World War II days, conscientious objectors were not very popular and that's what they were. I was the officer in charge of the camp, and it made me a little perturbed that these men had been sent here without my knowing their status. The fact they were C.O.'s didn't bother me in the least, but the reason given me, "We thought it best no one know that they were C.O.'s, or there might be trouble," bothered me a lot. The war ended and they disappeared as quietly as they had appeared.

Sal was the road crew foreman. His crew was unique. To my young fire crew, they were a bunch of old fuddies. Art was a school teacher about to retire from teaching, Unc was really old but full of good humor and from Missouri. Sal operated dozers, graders and all the equipment, as well as being foreman. He really knew roads and road equipment, and knew the forest road system like no one else.

This was the crew I had on my first fire. My fire crew had not yet been hired for the summer and when the fire was reported, I became pretty excited. Art was hard of hearing and didn't know what I told him, Unc was old and slow and kept telling me, "Just take it easy, it'll be there when we get there." And it was. Sal was pretty badly crippled from an old Forest Service injury. It took a while just waiting for him to get into the truck. At last, we were off. It was a railroad fire near Ash Creek, and burning hot. "Now don't get excited," Unc kept telling me as we neared the fire. I really didn't have much faith in my crew, or even in myself at the time, but things turned out well. The pump on the truck worked, and those old guys threw dirt and cut line pretty well. We had the fire controlled before

Cactus and Dutch arrived. I was on my way to becoming a fireman.

Art and Unc have long since died, but Sal is in his 20th year of retirement from the Forest Service and is still going strong. Recent hip operations have him getting around better than he did in 1945.

A FEW FIRES WORTH MENTIONING

Toward the end of my first fire season there was a hunter-caused fire on Black Fox. My crew had dwindled to two men, which seemed inadequate to me when we reached the fire. It was about two acres and moving up the hill faster than the two of us could handle. We hot-spotted, threw dirt, and cut brush for all we were worth, before I called the ranger station for help. It really hurt my pride to ask for help, the first time all year. Dutch said he would be there within an hour. We continued to work frantically until he came and were ready to drop. Dutch arived all right, with his deer rifle. "Where are the men?" I asked. Dutch lite his pipe, looked around, and said, "You've done a good job, sit down and take five." When we sat down, the fire quit spreading and things didn't look bad at all. Dutch knew all along that a fire in October in that type fuel would lie down about sunset and we would have the rest of the night to finish our control line. The first fall rain started that night, and we didn't even have to mop it up.

It was common practice to recruit hunters to fight fires in the 40's, which we did on the fire near Horse Peak. We had about ten reluctant men on the fireline, giving them instructions in how to build fireline and put out fire.

Roy, the Harris Springs fireman, was doing the talking, and his closing statement was, "There are about 200,000,000 acres of national forest land in the United States, about one acre for every person." "Then I'm going to let my acre burn," one grizzled hunter said, as he threw down his shovel and started to walk away. Everyone got a boost from this wisecrack, and all of us enthusiastically went to work on the fire, even the wise guy.

My first forest fire assignment was as a crew boss on the Lassen Forest, with Ivan, a top-notch fireman, as my sector boss. It was night, and my success in bossing a crew of 20 marine recruits was not too effective. In fact, the whole crew was asleep just outside the fireline when Ivan showed up. Admittedly, there was not much to do, as the fire was pretty cold in an area of burned out grass. But sleeping on the fireline is just not done. Ivan looked the situation over, and figured a way to shake and wake them up. He and I took fuses and set the grass afire around their sleeping spot, stepped back into the shadows and hollered, "Fire across the line!" Those marines jumped up, scratched line and fought fire like pros. Afterward, I convinced the whole bunch that an ember had gotten across the line and caused a spot fire which could have burned them up. There was no more trouble with sleeping marines that night. About 20 years later on talking to Ivan his comments were, "Now, Bob, you know that nobody in his right mind would do that dumb stunt!"

Two things made a fire on the Six Rivers Forest quite memorable. One was the transportation to get there. It was about 1950 or 1951, and we took off from Mott airport in a Ford tri-motor plane. They were pretty ancient even then, and this one was owned by a flying service in Missoula. It had been stripped of just about everything in the cabin, except for board seats to sit on, and had been converted to carry smokejumpers and dumping cargo by parachute onto fires. The door had been removed, leaving a gaping hole to watch the scenery go by. We flew at about 80 miles per hour, and only a few hundred feet over the mountain tops from Mott to Gasquet.

The airport in Gasquet was postage stamp size in the deep river canyon, but the tri-motor had no trouble sitting down on it. Kind of settled in like a big goose. It was one of the most fun flights of many I've taken in thirty-two years with the Forest Service.

On arrival at fire camp, things were as disorganized as only the Six-Rivers could be in those days. No one knew anything. Plans section was non-existent, the fire boss nowhere to be found and camp was set up in the biggest poison oak patch north of Shasta Lake. We bedded down for the night, getting rested for a big next day.

I was a crew boss with a crew of mill hands, assigned to fire out along a ridge top. Smoke was thick and no fire could be seen. The sector boss had the only map, and he started us working the fire line to fire the right hand side for about two miles. I told him it looked like there was more smoke to the left, and asked if he was sure which side to fire. He was sure. That is, for about 45 minutes. He came running back to tell us we were firing the wrong side. That 45 minutes was the only time during the day that the crew really worked. We had one-half mile of line fired by then, on the wrong side of the line. The rest of the shift was spent putting out what we had set.

At two p.m. the straw boss, one of the mill workers, told me they were going back to camp because their eight hours were up. I assured them that they weren't to go, and wouldn't go until relieved by the night crew some four hours from then. They did walk off the job, but they also walked off their mill job because they subsequently were fired by the lumber company.

I don't remember anything particularly noteworthy about one fire on the Angeles, except for the "john" at fire camp. The camp was at an old C.C.C. Camp on Baldy District and there was a row of toilets with horseshoe shaped seats, one side of which had been broken off from each seat. A scrawled sign above the toilets explained it all. "For the half-assed forest service officers in this camp."

One of my few assignments as maps and records officer was on the Monrovia Fire, Angeles Forest. It was a wintertime fire with one of the fire camps at about 6,000 foot elevation on Mt. Wilson. Cold, miserable and dusty, with Santa Ana winds blowing night and day.

The kitchen crew was unique. They were Krishna Laventa, a self-proclaimed Messiah, and his band of followers. He was an impressive rogue; handsome, tall, bearded, robed and barefoot. His group of men and women, young and old, were attired in robes of different colors, depending on their tenure and position within the organization. Probably, too, in relation to their relationship with Krishna, as he definitely had favorites among the fairer sex of followers. Several of these ladies shared his brightly colored, window draped, station wagon at various times day or night.

Each meal was preceded by the group singing a couple of verses beginning, "Blest be the tie that binds, our hearts in Krishna's love"—to the tune of the popular hymn, "Blest be the Tie That Binds." Since that time, it's been rather difficult for me to join in singing that tune at church without remembrance of old Krishna.

All of a sudden the winter hit us, with heavy snow starting during the middle of the night. Four inches of snow didn't faze the barefoot kitchen crew one bit. They cooked breakfast and paddled around like snow was an everyday occurence to them. They were good cooks, and a pretty interesting bunch of people, intelligent, and dedicated to their beliefs. There had been a little dissension in their ranks a few days before, if the rumor I heard was true. It seems that Krishna had taken a quite large sum of the group's money to Las Vegas and blown it over a crap table. Several months later, I read in the paper that Krishna had been killed by a blast from a homemade bomb. Seems he had been paying too much attention to the wife of one of the clan.

With the coming of the snow, we immediately started breaking camp. It was amazing that no one in camp knew how to put chains on vehicles, so I, being from the snow country, got the job. There must have been 25 trucks and pickups to chain up. An Indian on an Arizona fire crew helped me.

Several pickups were not chain equipped, and had to be towed out by other vehicles. I was driving one of these being towed with a 20 foot chain. Do you know what happens to the towee when the tower goes around a sharp inside curve on a narrow mountain road? You're right, you get pulled across the canyon! The driver's side was hanging in space, so I couldn't jump, but I hollered to my passenger to jump out. He didn't and by some miracle we managed to hang onto the edge with two wheels as our pickup bounced back onto the road. The driver of the towing truck hadn't even been aware of our predicament until it was over. He watched closer from then on. When I asked my passenger, the Indian who had helped me put chains on, why he hadn't jumped, he said, "Since you couldn't jump, I figured I'd ride it out with you."

Bohemotash (Bo-he'-mo-tash) had a nice name, but was a pretty nasty lightning fire. It was about 30 acres, and miles by trail from anywhere, so supplies had to be packed in by mules or dropped from planes. We tried both, and learned a lot. Don't do it.!

I went to the fire as sector boss, but ended up being a packer, with more mules than I could handle. Things went fairly smoothly from the trail head to within 100 yards of fire camp. It happened that the aircraft drop target was the same place that my pack string happened to be when the DC-3 unloaded its cargo. To make a long story short, an orange and white parachute settled gently over my lead mule. That was the last gentle episode for the next couple of hours. That mule didn't like that chute one bit and he whirled, jerking the lead rope from my hand. Then several other mules panicked as he crashed into them with a parachute flapping in the dust behind him. It looked like a riot at a rodeo, packs were flying, ropes breaking, heels kicking and packsaddles hanging under bellies. My saddle horse spooked and bucked a few times, but I managed to hang on by grabbing leather in a very undignified manner.

For two miles down the trail there were cans, bottles, bread, fresh meat, eggs, milk and whatever. Pieces of packsaddles, blankets, and pack gear were scattered and dangling in the brush—utter chaos and confusion until I started gathering mules. They had all stopped at a grassy meadow, grazing as if nothing had happened. It took hours to get the packs repaired and all the supplies reloaded. At least, we didn't have to worry about the airplanes, because it was about two a.m. when we reached camp again. Except for eggs, everything was salvaged from the disaster. Art, the regular packer took over at daylight and I slept in until six.

Modoc country burns fast and furiously. Sagebrush, juniper, grass and typically high winds make fires travel miles in short periods of time through the flats.

Ted and I had arrived at fire camp in late evening to be used on the next day shift. The fire boss showed us a place about 100 yards from camp where it was quiet to get some sleep. We must have slept well, because daylight came, we awoke, and the camp was gone, as was our vehicle. During the night, the fire made a run at the camp, so an evacuation was hurriedly made. Fortunately, it stopped short of camp and our sleeping area. We hitched a ride to the new camp, several miles away. The fire boss was very apologetic as he realized he'd forgotten the two men he so carefully accommodated the night before.

Lava bed country is one of the more interesting geological areas to me. After a small lightning fire near Six-Shooter Butte, my crew was hot, tired and thirsty. We knew of an ice cave near the fire so we decided to hike over to it to cool off and get a cool drink from the melting ice near the entrance. These caves are really lava tubes which have formed as molten lava cooled and solidified, leaving long tunnels throughout the flows. Not being a geologist I don't know whether or not my theories on how they formed are correct. Guess I'm entitled to a theory, though, so I assume the flowing lava cooled on top, forming a crust. Underneath, the lava was still molten, and flowed out from beneath the crust, leaving hollow tubes. It might have just shrunk as it cooled, leaving hollow places as it shrunk. You're entitled to your own ideas, so don't let me influence you. Back to the ice cave. In this particular cave, the floor is solid ice, several feet deep, with ice columns along the sides of the cave.

"Look!" shouted Bud, as we entered the cave. There were about eight cans of beer solidly imbedded in the ice, about four inches below the surface. Beer drinking was, and still is taboo during work hours, but under the circumstances I told the boys, "If you want to chip it out and have a can, it's O.K. with me." With much enthusiasm, they started chipping away around the cans. A half hour later, the first can was carefully removed from the ice. Bud swore, "It's empty, some son-of-a-gun put the can in upside down, with the holes down." The rest of the cans were empty, too, all carefuly placed so the open end was out of sight. "A sadist!" is how my crew described the man who had set us up.

Forest Service vehicles weren't always green. The last red tanker in Region V was at McCloud in 1945. It was no. 5, an Indiana by make. Big, bulky, and hard to drive, but a well constructed, brute of a machine, with huge tires which made it traverse the sandy flats of McCloud like an all wheel drive, which it wasn't. My only experience driving it was on a fire near Algoma, on the McCloud River. I was wheeling it along at a pretty good clip on a dirt road. As I approached a turn, I suddenly realized it wasn't going to make the turn and neither could I stop in time. Crash! I hit a stump head-on at about 15 miles per hour, expecting to crumple the bumper and grill, and possibly go flying through the windshield. I hardly slowed down! The stump uprooted and bounced into the brush, as the truck continued turning back onto the road. There was no mark on the bumper, except for a little paint which remained on the stump. The bumper on our present day tankers would have been wrapped around the stump like a pretzel, and the accident report would have been sixteen pages.

Porcupine Butte is lightning country. My crew and I were baby-sitting a two acre fire which we had controlled earlier in the day. We knew we would be there a couple of days, so we set-up camp and made some beds from fir boughs. Bud and Russ and I found a place together, but Rudy made his bed about 30 feet from us. We kidded him about being anti-social as we crawled into our sleeping bags that night. About dusk, we started hearing coyotes and they kept getting closer. Soon they were close enough to be seen standing in the shadows from our flickering campfire. We could see Rudy peering at them from his sleeping bag. I finally dozed off, but only long enough to be wakened by Rudy crowding his sleeping bag between Russ and me. He said it was too cold over where he'd been.

Tom, the radio technician, and I opened Black Fox Lookout one spring during a thunderstorm. After getting the lookout in operation, Tom and I headed for a fire we had discovered while at the lookout. It was at the western toe of Black Fox. There was nothing special about the fire, except that while we were working on it, the thunderstorm was still in progress. ZOT!!! A lightning bolt hit another snag about 75 feet from us. Slabs of wood rained around us, and the sound of the thunder and exploding tree was deafening. Lightning may not strike twice in the same place, but it can hit darn close.

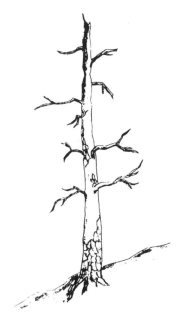

Snags—ecologists like them, firemen don't.

Los Padres country is famous for large fires, thick brush, some beautiful scenery, and a couple of unsavory creatures. Let me tell you about parajuelas (pa-hu'-a'-las) and tarantulas on the Cuyama Fire of the late 50's. I was cat-bossing on this particular fire, and my job was to lead two D-8 dozers down a brush ridge during the night in an area I'd never seen in daylight. The first Cat was to build a one blade width line, with the second Cat widening behind him. Armed with a typical, blinking, dim Forest Service headlamp, I was to guide the dozer down a ridge to the Cuyama River some two miles away. I won't say things were going great, because, even though we were building line at a fast clip, I was having nothing but troubles keeping ahead of the dozers in the 12 foot high dense brush. When I was close enough for the operators to see me, I was also close enough to be buried in the brush they were pushing. Some of the time I was waving my light for the operators to see, and the other times I was searching out a route to follow. Suddenly I looked down at my feet to see a gang of tarantulas waving their front two feet at me and showing the red insides of their mouths. Really there were only two, but sometimes two is a gang. Yes, I know tarantulas don't hurt people, but you just try convincing someone standing in a nest of them. We made it down the ridge by daybreak, but never used the finished fireline. A damp ocean breeze reached the area by dawn, and we held the fire on another ridge, thus saving thousands of acres of brush and tarantulas.

That night we bedded down on cots under some oak trees, not thinking about anything but some rest. About midnight someone hollered to move our beds from beneath the trees. It seems the parajuela tick, an ugly, rough-skinned, infectuous native of the area spends the days in oak trees and drops to the ground at night on unsuspecting firefighters and whatever else they attack. We each searched our sleeping bags and bodies, and sure enough, several of us found the insects, or whatever they are, in the beds. One man had been bitten and was taken to the hospital. The bite is not particularly painful, but usually becomes infected and takes months to heal, often leaving a nasty scar for years.

After getting back to sleep I was awakened by something big and dark nuzzling my sleeping bag. I came awake with a start, and flailed out with my fist. I had just punched a horse in the nose, and he snorted and galloped away through the sleeping men. When I woke up at daylight and mentioned the horse in our sleeping area, everyone looked quizzically at me. "What horse?" they asked. They were serious, and I had to find some tracks and more conclusive signs to convince them I hadn't been dreaming.

Some very insignificant happenings make certain fires memorable. The Nickel Fire on the Cleveland was not much of a fire by Southern California standards, but several things come to my mind about the Nickel. First, it started on the Leo Carillo ranch. For some of you younger readers, he was a quite famous western movie actor, usually playing the part of a rich Spanish landowner, and sometimes the "bad guy" in westerns of the 40's and 50's. Fire camp was in the ranch house area, and my remembrances of the huge Nubian goats and tumbling pigeons which entertained us during off-duty hours stick in my mind. Quite a few of the firemen chose to sleep in the big barn to escape the intense cold of the night, not realizing they were spreading their sleeping bags directly under the roosting rafters of the pigeons. By morning, the pigeon droppings had thoroughly camouflaged the sleeping bags, and a few direct hits had driven many of the sleeping men to the cold outdoors.

A few months later, the Stewart Fire burned completely around the Nickel. It was a devastating blaze, burning some 75,000 acres, and causing at least one death. On the sector adjoining mine, an Indian firefighter had been hit by a load of retardent as he stood on a rocky ridge, and the fall onto the rocks below had killed him. An incident like this really destroys the firefighters' morale, and many of the Indian crews were replaced on that fireline, along with the overhead.

On this same fire, during a long cold night shift, we had crews busily firing a tractor line about a mile from the nearest road when there appeared a group of young men and ladies from the Salvation Army, loaded down with backpacks of doughnuts, and dispensers of hot coffee. It was the best morale builder I ever saw. The men on the fireline took breaks from the hard work long enough to enjoy the refreshments, and went back to firing and holding the line with enthusiasm seldom displayed in the wee hours of pre-dawn.

It seems that at least half of my Southern California experiences were on the Angeles National Forest, a forest I know almost as well as I do the Shasta-Trinity. Woodwardia was a name to remember well. A bumper-to-bumper line of tankers and firefighters was attempting to fire out and hold a line below the Angeles Crest Highway above an advancing fire. Now this is not the safest procedure in the world, and I'm not condemning the action, because if the road were not held, thousands of man-hours work and thousands of more acres of valuable watershed would have gone up in flame. We were all aware of the dangers in attempting this strategy, and to my way of thinking, the risks were justified, with escape plans and adequate communication well laid out.

Just before dark, the fire got across the highway in one place, and took off up the hill into the rocks. Mac was the division boss, and I was one of his sector bosses. Because of approaching darkness and the steep, rocky terrain, the crews declined to chase the fire up the hill. Mac asked me if I'd take a crew up the hill and try to pick up the spot, which had probably reached 20 acres. We looked the situation over, and I suggested that Jerry from the Shasta Lake District, Hank, another crew boss and I go up and look at the situation from close range. It was dangerous from the standpoint of rolling rocks and hazardous footing, but the fire itself had laid down, and looked like the three of us could hot-spot and hold it to less than 30 acres during the night. We succeeded, and by morning it had practically gone out. The day shift crew mopped it up in a short time.

We felt good about our accomplishment during the night, and Mac wrote a letter of commendation for the three of us, which was really not deserved, because the job hadn't been that bad. We appreciated it, though.

The next day was different. After having a sense of accomplishment for the previous night's work, my sector was sent to mop-up a garbage dump at least a half-mile within the burn. It was smelly, smokey, and really didn't need mopping up, as it would have burned out by the time we could have put it out. The dump and I both smouldered all day. Me with the outrage of having to do that menial chore, and by the time I reached camp I felt obliged to express my feelings about wasting my talents and the men's time on such an unnecessary job.

Ed was fire boss and listened without much comment as I expressed my feelings. Don, also from the Shasta-Trinity listened to my outburst and egged me on from a distance. Then, as I finished my speech, Glenn, another sector boss from the Klamath lit into Ed about his equally unnecessary assignment for the day. Even I began to feel sorry for Ed, who was just doing his job, in a very cautious manner. The next morning we were on our way home, and I don't know whether the complaints from the Northern California overhead personnel was the cause of our being released or whether the job was completed to their satisfaction.

I've had the privilege, pleasure and opportunity of working with top rated firemen from all over California, both on regional fire teams and impromptu assemblies of highly trained and experienced men gathered together for specific fires.

A somewhat unique and effective system of rating for various fire assignments is used by the Forest Service. Firemen use a red card system of qualifying fire overhead, based on recommendations from the ranger district from which the fireman works, recommendations from the forest from which the fireman works, and from performance ratings from the various forests on which the fireman fights fires. Knowledge of fire behavior, on the job experience, training sessions with student evaluation, leadership qualities, and physical abilities, as well as a desire to fight fire are considered in making red card evaluations. As a result, the best fire peole are rated for various assignments, regardless of their regular employment title. Thus, a district ranger may wind up working on a fire for one of his own employees, who happens to have a higher qualification in fire control. Seldom have I seen any resentment shown in cases like this. The only inconsistency I've seen has been that some forests tend to rate their men higher or even lower than others—probably because their forest has fewer fires, and the men have less experience than those from other forests. It seems that the "fire forests" of the California region have better qualified and more accurately rated red card holders than those of the rest of the state. The fireman of the hot, explosive, Southern California forests are among the best I've encountered, with those of the Klamath and Shasta-Trinity forests in Northern California right up there with them. This is not to say that there aren't excellent firemen throughout the State of California, but the firemen from the hot forests just get a lot more practice.

Maybe it's the tendency for the men to homestead on my old district, the Mt. Shasta (Sacramento), that has made them among the best firemen in the region, whatever their job or assignment might be. There's Bob M., Tom, Ben and Jerry; any of whom could handle most any job on any fire. Others, with less off-forest experience, like Larry T., an excellent organizer; Larry M., an aggressive, effective fireman; Robert, Tom, Steve, Bill, Eddie, John and many experienced non-fire people, such as Paul and Bob B., gave the district a solid base of qualified firemen. These and others made my job easier during the seventeen years I was fire control officer at Mt. Shasta. My sincere thanks to them all—for knowledge, skills, untiring efforts, and especially for their friendship and understanding when things went badly.

Not only from their own forest or district, but from all over California there are memories brought to mind of men with whom you have planned, performed, and accomplished much in fighting hundreds of fires. Though I may never see them again, my friends from the national forests of California and the California Division of Forestry leave pleasant memories in the corners of my mind.

Also to be well remembered are those cooperators in fire control from private industry—particularly from the old McCloud River Lumber Company (presently Champion-International), International Paper of Weed; Kimberly-Clark of Mt. Shasta and Anderson; Cooper's Mill of Mt. Shasta; Trinco of Weaverville and numerous smaller mills throughout the Shasta-Trinity National Forest.

Porteous Logging, Crowe Brothers, Richardson's Contractors, Southern Pacific Land Co., Southern Pacific Railroad, McCloud River Railroad, the Castle Crag State Park, and the local fish and game wardens each played a great part in my work. My appreciation to these companies, men, and organizations is unbounded.

Last but not least, the cooperation, assistance, and goodwill created with the various fire departments from Weaverville, Dunsmuir, Castella, Mt. Shasta, McCloud, Fall River Mills, and Weed, has been a real source of pride and accomplishment for me. The sheriff's department, C.H.P. and local police departments, too, fall into this group. Each made my job a bit easier and more pleasant. Not that all was a rosy picture at all times. The Forest Service and they had differences, but we were always able to reconcile these differences in order to get a fire control job done.

LOST AND MISPLACED PERSONS

I'm sure Russ thinks it's funny now, but not back in 1946. We had a lightning fire about one-half mile from the Bartle Station, right on the flats, and surrounded within one-quarter mile by roads and a railroad. No trouble finding the fire. In fact, we drove the tanker within about 100 yards of it and walked on to the fire. It was getting dark, but we figured with a little maneuvering we could drive the truck right to the fire, so I told Russ to go bring the truck in. About an hour later there was no sign of Russ, and we couldn't hear the truck, so I sent one of the other men to see if anything was wrong. Within minutes he was back saying the truck was where we left it, but still no sign of Russ. We brought the truck on to the fire, wondering where Russ had gone. Maybe he had walked on over to the station for something. But another two hours went by and still no Russ. We called, honked the horn and blew the siren on the truck. Well, I was getting worried by then, so I left a man at the fire while Hal and I looked for Russ. It didn't take long to find him, because he was walking down the road toward Bartle, obviously tired, frustrated and mostly embarassed.

Here was his story. He couldn't find the truck because he hadn't paid much attention when walking to the fire. Consequently, he took the wrong direction. He had no trouble retracing his steps back to us and the fire, which he did several times. He was too embarassed to let us see him and admit he couldn't find the truck. So, time after time he headed back in a new direction to find it. Finally he really did get lost, from the fire, from the truck and from us. When he found the railroad track he followed it to a familiar road crossing from which he headed for Bartle. There we found him. The razzing from his fellow crewmen was pretty harsh, so I felt no reason to say anything until now.

It's really easy to get mixed up or lost in the forest, especially at night when everything looks alike. One evening we finished mopping up a lightning fire out in the lava beds and started back toward the truck. It was so dark that we used a compass to follow the route back. We figured a bearing of 270 degrees or due west, would take us where we wanted to go, but we also knew the truck was parked at the end of a dead-end road which also ran due west. If we missed the end of the road, our walking would parallel the road for about two miles before hitting another road running north and south. The hiking was pretty tough in the darkness, with brush and loose lava rock, so after going about long enough to reach where we thought the truck should be, our faith in the compass ran out. We bedded down about 11 p.m., hungry, and cold, to wait for daylight before looking for the road end.

Daylight came, and there right beside us, no more than 50 feet away, was our truck. Well, sleeping in the rocks makes for tough crews. We were the toughest.

One of the cardinal rules of crew bossing is to keep your crew together and in communication at all times. I learned a tough lesson by failing to do this.

I had picked up a ten man carpenter and construction crew at the nursery to take into the lavas to mop up a five acre lightning fire. The men were not trained firemen, and old Charlie was about 60 years old, not really eager to follow my bee-line compass bearing to the fire. We spent about three days on the fire under less than ideal conditions. We had no sleeping bags and subsisted on C-rations. Our physical and mental condition after three days was not good and Charlie was not about to follow me or anyone else across those lavas the way we had come in. There has to be a better way, he thought, and he took off on his own over my protests. During the first hour, we could see Charlie occasionally, meandering through the trees, picking his route. I very unwisely let him do his thing, as he seemed to be going in the right direction. It was about a two hour hike to the road, and we waited for Charlie for about an hour after we had gotten there. Finally, I sent the crew into town in their stakeside, while I waited for Charlie with my pickup and radio. Hours later Charlie still hadn't shown up, so I reluctantly radioed McCloud that I had a lost man. As the sun changed position, Charlie had changed his direction of travel, and instead of heading west toward the road, he had turned due south.

Ten hours and eight miles later he arrived at the McCloud railroad track, where he found a railroad telephone box and called the railroad company. They told us where he was and we went to pick him up on a railroad speeder. His shoe soles were completely worn through to his bleeding feet. The lava rock had really done a job on his shoes. He had taken cardboard from C-ration boxes to reinforce the soles. He was tired, hungry and very apologetic. I was relieved that he was all right. I got a well deserved lecture from Dutch and Paul on maintaining control of my crew.

Several years later, I was conducting a lookout training school. "Do you remember me?" one of the lookouts asked. It was Charlie. We had a good laugh over the "lava bed incident."

SEED CONE COLLECTING

Reforestation begins with seedlings, which get their start from seed collected in national forests. Most conifer seeds come from cones of various sizes and shapes. Sugar pine cones are sometimes twenty inches long and five inches in diameter, containing hundreds of seed, while an incense cedar seed is one of two in a small cone less than one inch long. In Northern California, most cone collections are from ponderosa, Jeffery, and sugar pine and Douglas fir, with a few white fir, red fir, and cedar cones collected. Collections are made by various methods, the easiest from mature trees felled by loggers at the right time of late summer or early fall. Care must be taken that the seed within the cones are ripe enough to harvest, and that the cones have not already opened, thus losing the seed.

In some areas the Forest Service has located and marked superior seed trees, which are harvested by Forest Service crews or contractors. This method assures a healthy, fast growing seedling, but also requires climbing by agile young crewmen. The usual procedure is to climb to the lower limbs with a ladder, and then use limbs as steps to reach cones at the top of the trees. A safety belt is often used around the trunk while the cone knocker uses a long handled stick to knock the cones to the ground. The stick is usually an old shovel handle, or McLeod tool handle with a thong attached around the knocker's wrist. Sometimes several gunny sacks of cones can be knocked from a heavy laden tree. Men on the ground pick the cones up, being careful not to get hit with a flying cone which might weigh over a pound.

We were knocking cones along Highway 89 in the early 50's, and managed to cause an irate motorist to threaten to take "the whole dumb crew on." One of the knockers hit a cone and it crashed through the windshied of a new Chrysler traveling about 50 miles per hour. The cone landed amid a shower of glass in the startled motorist's lap. He stopped to see what happened, thinking the cone had fallen from the tree, since he had no idea someone could be knocking the cones down. After telling him what happened, he got mad. Even after we salved his ruffled nerves, he still wasn't the happiest man alive. We in the crew pooled our pocket money and came up with $13, which the man accepted as payment in full. In later years, a full-blown, $200 investigation report, including a reprimand to the foreman (me) plus payment for the damages would have been made.

RECRUITING FIREFIGHTERS

Pickup firefighters were much more common during World War II than in later years of organized crews. We recruited men from anywhere. Sawmills and logging operations were the best source, but hunters and the man on the street would do. An agreement with sawmills worked well. If twenty men were needed, a call to the company would usually produce results. The request would be made to the mill superintendent who would first select the men whose absence would least affect the mill production. Or he might shut down one shift, and send the whole crew. The Forest Service tried to cooperate with industry by taking only what was needed and by releasing men and crews as soon as possible. Many mill workers and wood workers enjoyed the diversity of being on the fireline, though not all shared this enjoyment.

One weekend, the local mill wasn't running, so we were unable to get a crew from them. Dutch decided we'd better recruit from the street and from the McCloud Hotel, where most of the bachelor workers lived. We had signed up a few men, but the word must have gone faster than we had. When we knocked on one door in the hotel a voice answered, "I'm sick in bed." Dutch opened the door, and sure enough, the man was in bed. I was totally shocked when Dutch took hold of the covers and yanked them off. The sick man was fully clothed, including boots. It seems Dutch knew this fellow, and told him, "You fooled me last time, but you can't fool me twice." The "sick" logger grinned, and reported to the Forest Service as requested.

During the hunting season, Elmer, the woods boss, would gladly provide men for firefighting, but only if we guaranteed an equal number of hunters to match his men. Our system was to stop hunter vehicles along the road, and tell the hunters to report to the fire at a given point at a certain time. Their identity was assured by checking licenses, and I can't ever remember any hunter not showing up. We usually kept them only one shift, and if they had legitimate reasons, they were excused from duty. Medical excuses were checked through their doctors by telephone. Most of the hunters took their "induction" in good spirit. No, they weren't the best firefighters in the world, but that's the way it was.

Some Forest Service officers are empowered to recruit firefighters and enforce fire laws by being appointed as deputy fire wardens for the State of California. Most rangers, fire management officers and prevention technicians have these appointments, without which recruiting and enforcement are illegal.

Seldom does anyone question this authority, but one time I was glad I had it. There was a fire reported near Covington Mill on the east fork of Stewart Fork of the Trinity River. Larry at Trinity Center made initial attack, and nearly had it hooked when winds caused it to get away. I was coming from

Weaverville, and about one-quarter mile before reaching the fire there were two TD-24 dozers working a stripmining gold operation. When I asked them to bring their dozers to the fire the men declined, saying "We're making $60 per minute on this job." I was taken aback, but not for long. Knowing I had no authority to draft their equipment, and that I did have authority to draft the men, that's what I did. Well, most men would rather fight fire with dozers than with shovels, so they brought their dozers, and did an excellent job. I served as Cat boss, line locater, and swamper for the two tractors, guiding them as they built line through tall pole-sized timber, sometimes thinking the reason they worked so fast was to drop one of those trees on me. They built a mile of line in record time, and were released back to their "$60 per minute" gold mining operation in about three hours.

Three or four days later when I went by to pay them for their time and services, the men declined to accept payment for themselves or their tractors. Maybe they had been making $60 per minute.

HOW FOREST FIRES START

Contrary to Smokey Bear's statement that "nine out of ten forest fires are caused by people," in the West, and particularly on the Shasta-Trinity Forest, about one-half to three-quarters are caused by lightning. My estimate, from observing dozens of thunderstorms, is that about one in fifty downstrikes results in a fire. The other forty-nine are indeed hot enough to cause a fire, but because of several factors, don't ignite the object struck. A downpour of rain accompanying the lightning often puts any fire out. Sometimes the blast and accompanying wind with the strike will blow out any fire like it would a candle; other times the lightning may hit a rocky, non-flammable ridge, or even a succulent green tree which just won't burn.

When a dry, rotten snag is hit, it usually ignites from the intense heat of the lightning, and even a torrential thunder-shower won't quench the fire, as winds and dry wood feed the flames. This is typically how a lightning fire might start, though some smolder for hours before bursting into flame and enough smoke to be detected by lookouts or aerial reconnaissance.

There are literally dozens of ways that fires get started from people, with careless smokers being the greatest offenders. Smoking is such a way of life with some people that they're not conscious of it, thus making careless smoking an unconscious habit. Even rangers are guilty of careless smoking, with at least one fire known to have been started from a "timber beast." It was August, and a thundershower had wet down the area where a timber marking crew was working. During a smoke break, one of the guys sat down, carefully scraped away the top layer of litter, and smoked a cigarette. His intentions were good, but what he did was to scrape away the damp top layer, and uncover the dry middle layer of duff. This is what he ground the butt of his Kent cigarette into as the crew got up and left. Well, the butt obviously didn't go out, and the duff smoldered for a whole day before breaking out into a live fire. Since the fire was easy to control, it wasn't long before we discovered the Kent cigarette butt in the duff at the point of origin. We didn't know the Forest Service crew had been working there the day before, but when we reported the name of the fire, "Kent," and the location, one of the timber markers overheard the radio conversation where he was working, and knew immediately that he had accidentally started the fire. He hurriedly came back to the ranger station to see what had happened, readily admitting that he was the guilty one. He was billed for the suppression costs, as any guilty one would have been, and paid it promptly.

The Jug Fire was just that. A section crew from the McCloud River Railroad Company had lunch on an old punky log, and when they left to go back to work, their water jug, a one gallon clear glass bottle, was left on the log. Just as a magnifying glass will do, the jug of water concentrated the sun rays onto the punk, which smoldered and dropped hot sparks into the litter below the log. To prove our theory, we placed the jug on the log in a different place, and within minutes the pinpointed sun rays had the log smoking again.

One of the Castle Crag Park employees was a bottle collector, and found an old corked bottle filled with a liquid, and some solid chunks of something he couldn't identify. He uncorked the bottle and poured and shook the contents out as he walked slowly through the woods. Still searching for bottles, he turned around and saw several small fires where he had just walked. He didn't have any firefighting tools, so he headed for the nearest telephone to report the fires. The Forest Service and Castella Fire Department put out the fires, then started investigating the cause. It seems the bottle contained phosphorus chunks in water, and one of the properties of phosphorus is that it ignites on exposure to air. There's no telling how the bottle got where it was found, or why it was out in the middle of the forest. Needless to say, the park employee was not held responsible for the fire.

Kids don't deliberately set forest fires, but they sure do cause lots of them. The first one I can remember having anything to do with was not a typical "kid" fire. I was walking past the old Forest Service barn in McCloud, and caught a wisp of smoke coming out the door. There were two boys about eight years old stamping out a small fire in the loose hay. Sure, they had started it by smoking cigarettes picked up off the table at home. They were taking care of the fire all right, but when I came in they thought I needed taking care of too. One of them raised his trusty B.B. gun and aimed it squarely at my right eye. From his hip, too, mind you, never wavering as he said, "Don't you come any closer." Well, believe me, I didn't come any closer—fact is, I backed up a few steps as I tried to reason with the little outlaws. A truce finally prevailed, and he lowered the air rifle. We talked about the dangers of fires, smoking, and aiming guns at people. When I told him you should never aim a gun at anything you didn't intend to shoot, he replied, "I was gonna' shoot!"

We shook hands, made sure the fire was out, and they went away making promises to me which I'm sure were impossible to keep. Thus ended my first case of administrative action in a law enforcement case.

When I was a young boy, firecrackers and other *unsafe and insane* fireworks were a way of life for me and my friends, so I have empathy with the present generation of ten year olds who can't buy Roman Candles, Rockets, Whizzers, Torpedos, firecrackers, and any other fireworks not labeled "Safe and Sane." It was hard for me to get tough with the little kids who secured these contraband goodies from neighboring states, or from Uncle Joe back East. Anyway, nearly every 4th of July brought on at least one forest fire from fireworks used where they shouldn't have been. Usually it wasn't too difficult to find out who or how it got started, as the fires were near residences, and a few knocks on doors would reveal a family with fireworks, or a boy peeping around the corner of the house at the green pickup. Sometimes, even before any questions were asked, a tear-streaked, sobbing kid would appear, blurting out that he didn't mean to do it.

Whenever a group of say, a dozen kids are involved, it gets a little sticky in determining the guilty one. They sort of get into a circle and point to each other accusingly, or divide into groups and point accordingly. Then at other times, they'll all gang up on one poor little old kid with glasses, and make him the fall guy. I learned long ago to question them individually, so they won't know what the other ones said. Almost inevitably, the guilty ones will break down and confess, even though the confession may be, "My little brother did it."

Up until about four years ago, we had a policy of allowing

a kid one "mistake" before billing the parents for suppression costs, and so far as I can remember, there was never a second offense. However, the past few years have demanded that the Forest Service bill the child directly for any fire, with the assumption that the parents have insurance for this kind of waywardness in their children. Believe me, the parents don't cooperate nearly as well now as they did in the not too distant past.

Talking about treading on thin ice, that's when you've decided that little Don is a prime suspect in a fire case, but it so happens that his mother is the wife of the fire control officer, and also one of your wife's best friends.

"Say, Betty, where was Don when this little fire broke out across the street?" was the way Bob, the prevention technician approached my wife. Now, I don't know what the right approach should have been, but I do know that this wasn't it. He was informed in rather certain terms that he had been sitting right beside her at a football game and that she didn't like the implication of the question. As Bob backed away, he was assuring her that this was just a routine question asked all parents in the neighborhood and he was awfully glad that Don had not been involved. Even I felt a few aftershocks when I reached home sometime later. After all, Bob and I worked together.

One of the most frustrating "kid fires" that I ever had was the Power Fire at Junction City. It was a day after the fire started before we could pinpoint the origin, because the fire was about 10 acres at initial attack, and we just didn't take time to think about the cause. There was no doubt that the origin was just below a tree house in the backyard of one of the few residences at Junction City. There were candles, matches, and everything to associate with the fire, but one thing was lacking. It seemed that the family and children were out of town at the time of the fire, which was ascertained by neighbors. We never found out who started this fire, which burned three thousand acres and cost $90,000 to suppress.

Young people 14 to 18 years old usually don't produce many fires, but we had one at Baker Loop near Mt. Shasta which involved about 10 kids within this age group. A campfire was the culprit. It was a continuous, community type camp, with various users over a period of several days. We don't know who built the fire, nor who was the last to use it, but we did get a rather long list of admitted occupants of the camp. No doubt, someone had tried to put the fire out, as the ground was wet around the campfire except at one edge where they had failed to stir in the water. A telltale line of burned materials from the campfire to the adjacent litter showed how the fire escaped. Now, 14 to 18 year old boys and girls are a wise, close-mouthed group when there's a chance they're going to be in trouble. We finally narrowed this "probably most guilty" to two, after getting written statements from all of them. Several of the statements were classics, with statements like— "I don't know what you're hasseling me for, I wasn't the last one there, and anyway, I peed on it before I left." Another stated—"Why don't you ask Alice and Mary, they were the ones that were scared and wanted a big fire." A third, without being asked, reported—"I wasn't the one with the marijuana." We never collected from this one, but the land owner threatened to take them to court for trespassing and cutting green trees. Don't know how that came out.

Power line fires are not numerous, but some have been serious and expensive. One happened in the early 50's at Mott Airport. It was an early spring fire, and was caused from a mishap in McCloud, over the hill about 10 miles away. A house was being moved, and the gable end of the house hit an overhead power line. Not much happened there, except for a few sparks flying, but at Mott things were different. The shorted wires in McCloud resulted in a blown fuse box and transformer at Mott, which dropped hot metal into the brush beneath the pole. A strong north wind quickly spread the resulting fire over about 400 acres of brush. During the heat of battle on the fire, Ralph, the forest F.C.O. called a few of the bosses together. "Let's get where there's no one to bother us and make some plans to stop this thing." Ted, who had been working alone at the head of the fire since the start, replied with more than a bit of sarcasm, "Let's meet at the head, we sure as hell won't be bothered by anyone there." Ted was a tough, rough and talented firefighter who probably meant just what he said.

"Approach with Caution," is the byword in attacking power line fires. Even after the power company says "It's de-energized," it's good business to stay away from the innocent looking wire hanging limply from the cross-arm or on the ground. Lots of things cause powerline fires—snags or trees falling across the wires, poles snap off, insulators pull out, cross-arms break, fuses melt, and sometimes the wires just break for no apparent reason. A new cause showed up in 1976. One method of stringing a new line across a canyon is to propel a missle across a canyon with a mortar type gun. A line is attached to the projectile, which is then used to pull the powerline wire across from the other side. This unique system has been used several years, according to Mr. S. of the power company, but on this occasion, the wadding in the mortar shell caught fire and ignited the brush where it landed some 40 feet down the canyon. Fortunately, the fire was suppressed quickly by the power company before we reached the scene, but the incident pointed out a new way for fires to start.

FALSE ALARMS

Nothing embarrasses a lookout more than reporting what turns out to be a false alarm. It happens to the best and the worst. Let me tell you about a few.

My first was a tall spray of water from a pipeline. There was a 24 inch redwood pipeline running from Lakin Dam on the McCloud River to the town of McCloud. It lay on top of the ground and occasionally sprung leaks, sending sprays of water high into the air. From Grizzly Peak with the sun in the right position, these leaks looked like a plume of smoke. Within minutes, Cactus had found the "smoke" and told me what it was.

Another false alarm I reported was water vapor rising from a natural hot spring near Big Bend on the Pit River. This happened early one cold morning. Glen, the Big Bend fireman quickly discovered what my smoke was, and apologized for not informing me of this condition. Evidently this had happened in previous summers to lookouts at Grizzly and Bunchgrass.

Right after World War II, a disabled veteran was hired as a lookout on Bunchgrass. His disability was both physical and psychological. When he reported a smoke one hot afternoon, a full blown dispatch of crews and equipment was made. No one could find the fire, which the lookout insisted was there. A recon plane was dispatched to get a better location on the fire. The recon couldn't find it either, even by flying from directly over the lookout on the azimuth reading from Bunchgrass. Finally the recon observer decided the smoke had been a reflection of the sun on a large buckskin snag, which was in the location given by the lookout. By this time the lookout was embarrassed, befuddled, and very emotionally upset to think he had caused so much trouble and expense by reporting what proved to be a false alarm. No amount of reassurance by the dispatcher could calm him down, so the fire control officer at Fall River Mills, fearing that the man might do something drastic, hurriedly drove to Bunchgrass and brought him down. Wisely so, as the lookout told him he seriously was contemplating suicide by hanging himself when the Forest Service officer got there.

Even Leila, a most dependable lookout, made a mistake. The Sheep Well fire was to the north of her and at two a.m. she reported a fire, with flames visible near the top of Harris Mountain, to the south of her. By the time we were dressed and in the truck, she called back. "Bob, I was looking at a reflection of the Sheep Well fire in my south window, and didn't realize it until I went outside to get a better look."

On the same date and hour in two consecutive years, Sims Lookout reported a smoke on the bank of the Sacramento River, drifting across the river. Each time there was a different lookout, but the sun position on that date and hour causes a reflection on the moving water which both lookouts took to be smoke.

The Aurora Borealis on several occasions has caused local people to report "a glow from a large fire in the north." This is not common though, as the Aurora seems to occur at intervals of several years.

Even the lookouts have been fooled by the glow of the rising moon coming from behind a timbered ridge when the air is smoky or hazy, though they quickly realize what they're seeing as the moon appears from behind the trees.

At least twice have I checked a lookout's report of a small fire in the trees or brush, only to find a young couple parked on a secluded road, with their tail lights glowing.

During hunting season, it takes an experienced lookout to determine whether a smoke is from a campfire or the beginning of a wild fire. Consequently, many false alarms happen during hunting season. These often result in expensive dispatch operations, including aircraft, because all smokes are assumed to be wild fires until proven differently.

HOW TO FIND FIRES

Most man-caused fires are easy to find because they're on roads or in places of easy access, but a typical lightning fire can be a problem. Especially difficult are those in dense forests on flat ground. Before departing for a lightning fire, the fireman must know the location determined by lookout azimuth readings on the dispatch map, taking care to know which side of the hill, mountain, or creek the fire is on. The dispatcher should have all this information ready for the fireman. If possible, the fireman must be equipped with all necessary tools to find the fire. These include a radio, map, protractor, signal mirror, and flashlight. When a lookout says the smoke is faint, or intermittent, it is essential that the fireman knows exactly where he is, and the direction of travel he is to take. Topography often determines the best route, which may not be the shortest distance.

Typically, the fireman will leave his vehicle at a predetermined location on the map and road, taking a compass bearing as determined by the protractor and map. He will walk on this bearing for a predetermined time or estimated distance. Usually this will lead him to the fire. If by then he has not found the fire, he will find a high point to look over the area. On flat ground he may have to climb a tree. From these points, if the fire is not sighted, he possibly can see a lookout who can help him find the fire. A signal mirror flash will show the lookout where the fireman is in relation to the fire location. With this information, the lookout tells the fireman the direction to travel and approximate distance. This distance can be closely estimated by determining the difference between the azimuth readings on the fire and the fireman's location.

The fireman can pass within a few feet and miss seeing a wisp of smoke in the top of a green tree. Naturally, the larger the fire the easier it is to find, but also the more difficult to suppress. In the forties and early fifties, airplanes were more of a luxury in assisting the fireman in locating small fires, but even then, a system was developed by Merv to help the fireman. Recon planes were used after lightning storms to assist lookouts in locating fires. There are about 300 feet of toilet paper on a roll (bet you didn't know that), and by folding about a five foot long tail into the palm of your hand, the roll can be tossed out the window of a plane, unrolling into a 300 foot very visible white ribbon. It drapes itself over trees and flutters in the breeze for hours before breaking up. A little practice and a toilet paper bomber can mark an area around a fire that no one can miss. It really works, and is quite effective.

Sometimes a recon plane will "talk" a crew into a fire by telling them the direction to travel, and the easiest route in. This can be rather expensive, because a crew on the ground travels slowly, and recon planes get more expensive each year. The plane can get the men headed in the right direction, then leave for other work, to return only if the crew still fails to find the fire.

A well trained lookout can be as much help as a recon plane where fires are unobstructed from their view. At night, lookouts can guide vehicles and foot troops to fires by observing headlamps and car lights. Whether it is day or night, the job is harder on the ground than it appears to be from the lookout or airplane.

A GREEN MONSTER

It was in 1946 that the McCloud District acquired a tank truck unique in the annals of Forest Service history. Leastwise, I never knew of another one like it. The motor was a Ford V-8, vastly underpowered for the size and weight of the truck, which was a six-wheeled, all wheel drive Ford chassis. Behind the cab was a three-man jump seat, then a 500 gallon tank, followed by a Wisconsin V-4 engine and four stage centrifugal pump. It could go anywhere, especially with Hal driving it. He was a fine driver and pump man, and he babied that machine like a solid gold Cadillac. There was no rushing Hal, or the truck, but wherever you wanted that truck, he would get it there. The gear ratios were such that at highway speeds of thirty miles per hour, the engine was reved up to ninety, and to go over the McCloud Summit, it just took patience for about a half-hour for the truck to travel the six miles. But, boy could it go down the other side. It cornered like a Maserati while making up for lost time. Because of its slow speed, the crew traveled in a pickup to fires, and used the tanker whenever it got there. We tried to have a control line around our fires by the time the 6 x 6 got there, and this served as an incentive for the hand crew on every fire.

COPTERS—WAY BACK WHEN—AND NOW

With the ending of World War II, helicopters became a tool of the Forest Service. I believe it was 1947 that the Shasta Forest first made use of these amazing machines. Fred was the contractor on the Shasta, and had a small copter based at Mott. As I remember, most of the use was in training and familiarizing fire control personnel in safe use of the machine. Men were loaded and unloaded, tools were lashed into place, and light cargo packed and hauled to lookouts. A few fires were manned by landing on nearby ridges, and hiking short distances. I've read that the first jumping of men from copters onto fires occurred in Southern California about 1950. These may have been the first sanctioned jumps by Forest Service personnel, but actually, there were jumps made directly onto fires in 1947 on the McCloud and Redding Ranger Districts.

Bud, a young 17 year old crewman on the Sacramento District and at present the Sierra Forest F.M.O. was one of the first to jump. Fortunately, he wasn't killed as he plunged through an oak canopy, which was thought to be low manzanita. Now, this was wearing no more than Levis and tan shirt, with no protective clothing. I know, because that's the way Joe and I jumped into a small fire near Bald Knob west of Grizzly Peak a day or two later.

Because the copter lacked the power to hover as they do now, our instructions from the pilot were to "jump when I tell you to." This was to be done when the copter was closest to the ground, and before it started to climb, or before the hill started to fall away beneath us. Well, all of us survived the jumps, but it was decided that this practice should be discontinued. We'd go back to the system of landing men at the closest safe helispot and walking to the fires. This was a good decision, as I'm sure someone would have been killed had the practice continued. Several years later, with more powerful helicopters and padded jumpsuits, the art of helijumping was resumed, but even today it is not used much in the California Region. I believe it will again be done in California with highly trained personnel, and also with the used of let-down cables to lower men into timber areas, which is now being accomplished in other western regions of the Forest Service.

The first departure from the small underpowered copters came with the coming of the "Alouette," a hot, fast, French built machine, several of which were privately owned and under contract to the Forest Service. One of these was owned and operated by Bob, working out of Fort Jones, who did much flying on both the Klamath and Shasta-Trinity Forests. Bob was a skillful pilot with an exceptional machine, and flew with what I felt was an attitude of daring adventure, totally confident of his ability as a pilot. In talking to him one day, he pointed out that his machine was the best available, with more than adequate power, and maneuverability unmatched by other copters. The fact that he seemed to enjoy skimming over the ridges, and treetops was explained in his statement, "Bob, the Alouette can and should outperform those other small copters by a two to one margin, and so far as safety goes, if this motor fails, the result is the same whether we're flying high or low, fast or slow, since the Alouette doesn't have autorotation capabilities." (Autorotation is the ability of a copter to fall straight down gently, without power, except for the movement of the top rotor windmill action as it drops through the air.)

Well, I flew with Bob on numerous occasions like the

Rarie Fire at Weaverville, where we operated from a narrow stream bed surrounded by trees. Real skill was required to land and take off. He had it.

Bob and his Alouette were on the Basin Fire of the Sierra, doing a tremendous job shuttling men and equipment, plus laying hose and placing relay tanks and pumps. The weather was so hot that between flights, Bob would wade into the Kings River with his clothes on, come out dripping from head to foot, and take off with another load.

Toward the end of the fire we were removing a hoselay from the steep slope alongside some penstocks above one of the Kings River power houses, using the Alouette to lift out the relay tanks and pumps. Bob was hovering over the Pacific pump, with rotor blades only a few feet from the rocky slope, and I was lifting a pump onto the cargo racks atop the skid on the downhill side of the chopper, when, "Zap!" a charge of built-up static electricity jumped from the copter to the pump I was holding about chest high. I tumbled down the slope with the pump all over me.

About twenty feet and several bruises later I regained my feet and composure long enough to appraise the situation. The copter had circled around and Bob signaled me to try again with the pump. This time, there was no "zap" as I gingerly set the pump in the rack. Later Bob told me, "That just happens sometimes, but usually not once after another." The flash really didn't hurt, but scared the heck out of me. I don't know yet whether the electrical arc knocked me over, or if I just jumped backward from a reflex action. Anyway, no damage done, but things like that can cause accidents. Just lucky, I guess.

A few days later, and a few hundred miles to the north, I worked with Bob again. This time on my own forest. The Hatchet Fire was only a hundred acres or so, near Trinity Center. As line boss, I was getting a birdseye view of the fire from Bob's Alouette. He asked me, "Do you hear that vibrating noise toward the rear?" I could really not tell it from other vibrations and noises, but Bob's trained ear could pick up something that I couldn't. "It's in the tail rotor bearings, and

I've got to get it fixed as soon as this fire is over," he told me, as I tried to hear the noise or feel the vibration. "If you think it's serious, let's set this thing down until it's fixed," I responded, but with his assurance that no serious consequences would happen I forgot about it as we continued to fly.

The next morning, Bob had a previous charter for his ship up in the Marble Mountains, and left the fire, telling us he would be back within a few hours to finish our work on the fire.

He never returned, and late in the afternoon, the wreckage of his copter with his body was spotted near the point of pickup of his charter. Several years later, a deposition was obtained from me by a law enforcement officer to be used in a case involving Mrs. Bob, the Alouette manufacturers, and an insurance company. I never heard the final outcome of the case.

The impression I have of Bob will never be erased with time, and I'm sure others associated with him feel likewise. Despite an air of brashness, self-confidence or maybe even self-esteem, I'll always be sure that it was not pilot error which caused the crash. He was the best, and he knew it. If pilot error could include failure to recognize the seriousness of what appeared to him to be a minor mechanical deficiency, he might be guilty.

But for the pre-scheduled charter flight of about two hours, I could have well been in the copter when the machine failed.

Probably no use of helicopters is more hazardous than in forest fire control work. I never really felt totally safe while flying, but of necessity in my fire line jobs, many hours were spent in and around helicopters. Witnessing two accidents, one with fatalities, and one a near miss, plus having two close friends die in helicopter crashes no doubt has affected my confidence in them. I can't condemn the machines so much as the unreasonable demands we put on their performance, and on the skills of the operators. At the same time that I have some reservations toward helicopter use, I also recognize the huge contributions they have made to more effective fire control operations. Their speed and efficiency in transporting men and equipment saves thousands of man hours each year, and increases the efficiency of troops by several times in morale, time and fatigue factors.

I look back at a helicopter mission on the Canyon Fire of the Angeles which was successful, but in retrospect, know that it should not have been attempted. A hot-shot crewman injured an ankle while building fireline on a very steep slope about a mile from the road. There was no place to land, because the slope was so steep that the main rotor would hit the ground long before the skids touched. Our solution was to build a shelf on the hillside, which was no easy job, but unlike most Angeles fires, there were several small trees growing nearby, which we felled, cut to length, and built a crib for the copter to land on. Even with the completed shelf, the pilot had to set the chopper down on one skid, with the rotor only missing the hillside by a few feet, as he applied power to keep the machine level. We loaded the injured man into the bubble and the copter quickly fell away into the canyon with his passenger, heading for the hospital. When we found out his ankle was not broken, I'm afraid I was a little disappointed, as I was basing the need of rescue by helicopter on that assumption. About an hour after this rescue, I witnessed another copter crash into the brush on the adjacent sector, which was the second fatal crackup on this fire. A total of eleven men died in the two mishaps.

My friend Sam, died in a routine flight in the high country of the Inyo, when the copter he was in crashed, in the late 50's. He and I had worked together and he had lived in my home in Fall River Mills before he married. His death was a blow to me and my family, as well as all those who knew him. He had married and had a little girl, who, the last time I saw her, was a replica of Sam. Sharon, his wife, has since remarried, and still corresponds with us at Christmas time.

An incident on the Shasta Fire in 1967 was unique, and rather disturbing to Bert, the pilot, and Bob. It was early morning in mid-October, and colder than the anatomy of a witch. The heliport was smogged in from the smoke inversion of the fire, but with enough visibility to take off. Bob was to take a look at the fire perimeter for Lee, the fire boss, and report the early morning conditions. Because of the cold, the doors were in place on the copter bubble. Well, the smog had acted like smudge in an orchard, keeping the ground surface slightly above freezing, but when the copter rose above the inversion layer, instantly the bubble glazed over with ice and frost on contact with the sub-freezing air above. There they were, about 100 feet above the trees, with no visibility. This was a new experience for both Bert and Bob, but with Bert's skill, and Bob's directions by peering out his opened door, the ship was brought safely back to the heliport. It took a couple of cups of coffee to get them back in the proper frame of mind to fly again, and they were very short with Lee when he asked what were they doing back in camp when he had sent them on an important mission. Lee thought it was much funnier than they did as the situation was explained to him.

Once I got into a little trouble over a helicopter, and admittedly I was in the wrong, but I believe a few dollars were saved in the process.

Roy was a lookout-fireman on Backbone Ridge, when he reported a small fire from a dry lightning storm about three miles from his station, in some flashy type fuels. As I was deciding how to handle the fire, I saw a Forest Service copter flying up the drainage and flagged him down by radio. The pilot said he was going to Grizzly Creek to pick up some jumpers from another fire. This seemed like a lower priority than my new fire so I told him to pick Roy up at Backbone and put him on the ridge just above this new fire, which he did.

Now there's really nothing wrong with that, except the forest dispatcher, Bud, should have been told and allowed to make the decision. I really intended to do so, but communications weren't too good from my location, and in the excitement I somehow forgot to notify Bud. After about an hour, with the fire manned, and the helicopter delayed an equal time, Bud started checking on its whereabouts. About third hand, he found out I'd "highjacked" his copter and not told him. The lecture I got from the dispatch office and Bob, the Forest fire control officer were pretty pointed, and it was probably several hours before I made another goof. Roy said the action probably saved a few acres, as he had a nip "n" tuck battle in controlling the fire. "Thank's, Roy, I needed that!"

The years have brought on bigger, better, faster, and more powerful copters, increasing their potential beyond man's dream of just a few short years ago. Besides having copters capable of hauling crews, instead of just one or two men, now the Forest Service contracts jobs requiring lifting capacity of 20,000 pounds. In 1975 we moved Black Butte Lookout about 40 miles airline to Hogback on the Shasta Lake District. The pre-fab metal structure was picked up bodily with a Sikorsky Sky Crane and gently placed at Hogback without so much as a cracked window. It was not easy though, at 6,300 feet elevation, the sky crane had to burn off several hundred gallons of fuel before the lift-off of the 10 ton structure. This necessitated landing in the plantations below the lookout, refueling the ship, and then flying to Hogback in the denser air at lower elevations. The old Black Butte lookout was put into service at Hogback in 1976, and the lookout had never had such a year reporting fires. The severe season in early 1976 resulted in fires within her view almost daily, and she kept the air busy with "C.D.F. Fire Traffic."

Helicopter logging came into its own on the Shasta-Trinity during the mid-seventies, with Columbia helicopters rapidly and efficiently logging millions of board feet of timber on the Mt. Shasta District. Results of the logging will probably greatly affect future logging practices both in federal and private logging operations. Little or no visible scars show from the logging, making ecologists and environmentalists happy, though it might not help the spiraling cost of lumber.

TRAIL WORK

The McCloud River, below the McCloud Dam, is still one of the wildest, and most beautiful areas in northern California, or anywhere else, for that matter. Steep canyon walls towering above the river are covered with tall timber and dense undergrowth. The river itself tumbles and twists over rocks and around bends. White water turns into deep blue pools before speeding up again over riffles and rapids. Huge ferns, big leaf maple, and other water loving plants grow in the rocks and shallow edges of the river. Before the dam was constructed in the 1960's, the wild, primitive condition existed all the way up river to the Hearst Estate. It was in this setting that I worked on forest trails in the mid-forties. Sometimes Cactus and I would go down and camp for a day or two at Ah-di-na or on Hawkins Creek. The trails were in pretty fair shape then, so we could do two to three miles of maintenance in a day, working both directions from our camp. On some occasions I would pack a two man crew into an area and leave them for a week, then move them again the next week. It was these pack trips that I liked. Just like vacations, they were.

One summer we hired a regular packer, a young man called Tex, who wasn't a bad packer, but his rendition of "My Adobe Hacienda" was pretty bad, and he sang it most of the daylight hours. When he wasn't singing, he was telling me how good a shot he was with a pistol. In those days, most all rangers carried a pistol or rifle, not for protection, but just to use on porcupines, or target shooting. (We did kill a few deer during buck season, which was O.K. so long as we took a couple of hours of annual leave to take care of the deer.) Tex was riding about 20 feet ahead of me one day when a big green rattlesnake slithered onto the trail between our horses. Tex turned his horse around to look at the snake just as I drew my .38 from its holster, and put a slug through the snake's head at about 20 feet from the top of my fidgety mount. No one was more surprised than I, as I tucked the pistol back in the holster. I must not have let my surprise show, as Tex never mentioned his shooting skills again. He just kept on singing.

In 1954, I was in charge of trails on the Pit District, though Sid did most of the inspection, and John was the trail crew foreman. There were many miles of trail in the Big Bend area, as this country was not opened to extensive logging until about this time. John supervised and worked with two crews of two or three young men, and really managed to get good production. He was old and wise, to the young men, as well as having a flair for keeping their morale up. His sense of humor was different, to say the least. I spent a few nights with the crew camped on the trail. While the crew was working near camp, John went back to camp to prepare for supper. Part of his preparation was to carefully take the labels from dog food cans and put them on cans of beef stew.

"Hey, Joe," he said, as he started preparing supper, "hand me a couple of those cans of dog food, I think we'll try it tonight." "Yeah, John, let's have some," Joe joked back, not to be taken in by John's foolishness.

Joe and Tom both turned a little lighter as John emptied the cans into the pan. "Now, that doesn't look bad, does it?" Both Tom and Joe were now assured that supper was to be dog food, which John had convinced them, was perfectly good food for humans. After John and I took a hearty portion on our plates and started eating, they took a cautious bite, but never really ate as hungrily as a couple of hard working young bucks should have eaten. I don't know yet whether John ever told them what he'd done.

WINDMILLS AND WELLS

It wasn't until after we built the tower for the windmill that I found out the well was yet to be dug. We prefabricated the tower at McCloud with this job plan. "Bob, I want a 20 foot tall, four legged tower built to mount this windmill on. We'll take it to the site when you're done." No problem, as we had enough treated telephone poles on hand to serve as legs, and a few two by eight planks for diagonal bracing. After drilling a few holes for bolts we assembled it, to make sure everything fit, then disassembled it to haul to Toad Lake. There's more than one Toad Lake in the country, but this one is really just a clear area surrounded by lodgepole forests, which has a puddle in the middle in the spring. When we got there, I expected to see a well where we were to put the pump.

"Didn't I tell you," said Dutch, "You also have to dig a 20 foot well." He knew we would hit a water table before that depth. Digging was really quite easy. In no time we had gone 4 feet through the sand and soft pumice gravel. We then placed a four foot length of 48 inch culvert on end into the hole, and we had to throw the dirt out with a shovel, while the rest of the crew hauled the dirt away. As digging progressed, a second piece of culvert was placed over the first, to form a continuous cribbing of the walls. By digging underneath the bottom edge of the bottom culvert, the weight of the culvert made it drop further into the hole, and each section above also dropped into place. It was a good plan, and it worked. The deeper the hole got, the slower it went, as the dirt had to be hauled up in buckets. Nevertheless, in a few days we had completed the job. Damp dirt gave way to mud at about 12 feet, and at 16 feet we had more water than we could bail out.

The windmill, too, was a success, but about two years later, the sheepman, for whom the well was built, quit leasing the range. He said the Forest Service had ruined the range by putting out all the fires, and there was no more good browsing for his sheep. That was the same year we had a ten acre fire near Harris Mountain just a few hours after a band of sheep went through. We never were able to prove the sheepman started the fire, but the herder said it could have started off his horses shoes striking a rock. I doubt it.

McCLOUD BRIDGE—NOSONI CREEK

Working the winter of 1945-46 should have been the end of my Forest Service career, but I hung in there. It wasn't the work that was bad, but I caught poison oak so severely that it took about an hour each morning to get the swelling around my eyes reduced to the point I could even see. Total misery for several weeks, and I didn't even bother to see a doctor. Others had it, too, but I'm sure I was the worst. For a single man, the job itself wasn't bad. We were surveying for a road from McCloud Bridge to Nosoni Creek, and up the creek, over to Fender Ferry on the Pit River. The survey crew consisted of Ted, the assistant to the forest engineer, Bob M., Roger, Butch, and me. Ted was a dedicated, hardworking man, who got much from his crew, not particularly by being a good leader, but by being a good example. Quitting time to Ted was when he could no longer see the rod or range pole through the transit or level. Time meant nothing to Ted. He would take the hint to go back to camp if one of us held a match alongside the rod, indicating it was getting dark.

Now, this job was not easy. We ran center lines, profiles, clearing lines and cross sections on steep rocky slopes through poison oak like no one except Shasta Lakers have ever seen. Bob M. and I worked pretty hard, because we didn't want that old guy, Ted, to show us up. Roger would rather be doing something else, and it showed. We just had to prod him along on occasion. Butch, from Scott Valley, was something else. A hard drinking, hardworking, back-country type he was. One of the last of his breed. Along with this, a natural comedian with a slight speech impediment and one eye which didn't quite track with the other.

One time Ted hollered at Butch, "The rod is upside down!" to which Butch bent over, with his head down looking at the rod and replied, "It 'ook aw 'ite to me." Ted's sense of humor did not always follow the rest of the crews, who really enjoyed Butch's comedy style.

Our camp consisted of tents and tent frames with three men to a tent, plus a large mess hall tent with a cook, who wasn't the greatest.

Most of the crew cut brush and trees in the road right-of-way, piling and burning it as weather permitted.

The country was and still is beautiful above the McCloud River arm of Shasta Lake, and up the Nosoni Creek drainage. Limestone outcroppings and formations loom out of the dense vegetation. On our time off, some of us explored Samwel Cave near the new road, later to become more well known to geologists and scientists, as well as the general public. Visitor tours are now conducted by the United States Forest Service.

LEARNING TO SKI

I won't say I learned to ski in 1945, because in 1976 I was still trying to correct all the self-taught "wrong-way-to-do-it" techniques I'd perfected over the years. My first attempt at skiing was like this. When Dutch and Cactus asked me how well I could ski, I had to admit I'd never seen any skis close up. "You had better get acquainted with these pretty quick then, because tomorrow morning we'll be taking a snow survey," Cactus tells me. There were such things as ski boots back then, but Cactus never bothered with such frills. A couple of horseshoe nails driven into the high heels of my logger's boots served to hold the cables in place, while my toes were securely held in the bear-trap bindings. It was really a pretty good setup. Your skis seldom came off, and it must have developed some pretty strong ankles. For the next two hours, I practiced on the hill across from the Forest Service in McCloud, just climbing up and sliding straight back down. Pretty fun, I thought, as I prepared for the next day.

We started at daylight. The snow was crusted and easy to walk on. The terrain from Highway 89 to the Ash Creek cabin is a long gentle slope of about 13 miles; a good way to get used to cross country skiing, as no particular skiing skills were needed. Later in the day we got into softer snow where we took turns breaking trail, and where our 25 pound packs seemed noticeably heavier. Cactus was strong and durable, never seeming to tire. We reached the cabin in good time and in good shape, except that my leg muscles ached from this new form of exercise. A quick fire warmed up the snow-covered cabin in no time and a good hot meal renewed our energy. Afterward, we measured the snow on the Ash Creek Course which is just a stone's throw from the cabin.

Neither of us slept too well, because the old kapok fill sleeping bags were damp and musty smelling.

The next morning, snow was falling, forming a soft blanket over the old snow. It was beautiful as we left the cabin toward the Brewer Creek Course some six miles away. The new snow kept piling up, and long before reaching Brewer Creek we were fighting through knee deep fresh snow, coming down thick and heavy. We finally reached the course by following markers along the route. With the heavy snow and poor visibility it would have been impossible to find the course without the markers.

By the time we got back to the cabin, it was about 3 p.m. and we were cold and I was tired. But Cactus thought about those damp sleeping bags, and said, "How do you feel about going back to town right now?" Well, those sleeping bags weren't all that bad to me, but not wanting to show any lack of strength, courage, character or whatever virtue might be missing, I said, "Let's go." At least it was all downhill for the next 13 miles. Really, it wasn't too bad. Our old tracks helped make the skis go faster, and we made good time, though my shoulders ached from the pack straps, and I was *tired!* About one mile from the highway we sat down to rest. It was pitch dark by now and we set our packs down while we rested. When we got ready to go I took off without my pack. So help me, I could still feel those straps biting into my shoulders all the way to the highway. When I got there, no pack. I was speechless, disgusted, and couldn't believe it. Cactus wasn't speechless. "You cabbage head, how could you forget your pack?" Plus a few other choice words and phrases. Twenty-five miles, plus two, equals 27 for the day. For my first two days on skis, I was proud to say that I survived.

COOL, CLEAR WATER

The most valuable product of the National Forests of California is not timber, but the winter blanket of snow covering the millions of acres of high country. Not only does it provide protection from freezing to young trees and vegetation, it provides moisture in the spring and into the summer for our timberlands. This is only the beginning of its value to Californians and the rest of the United States, which gets 40 percent of its foodstuff from California orchards and farms. Water from these snowfields affects everyone. Skiers and snowmobilers use it as it is. Fishermen, swimmers, boaters, waterskiers and miners use it as it melts and flows downstream channeling into large reservoirs behind huge dams. Federal and private power companies use the water as it flows through turbines, converting water power to electrical energy. From here the water is directed to irrigation districts for further distribution to farmers through a network of canals and ditches throughout the great agricultural areas of the central valleys of California. And there is still water to flow to the Pacific, keeping salt water from intruding into the rich delta regions.

To properly manage this unpredictable resource, the State of California Division of Water Resources, with the cooperation of the United States Forest Service, Bureau of Land Management and private industry, has set up hundreds of checkpoints throughout the headwater drainages of all major rivers. These checkpoints, called snow courses, are used to gather data on snow depth, water content, density percentage and other pertinent information. From mid-October until April the snowpack deepens and compacts into a dense blanket, sometimes reaching 20 feet or more in depth at 7,000 foot elevations in the Mt. Shasta, Mt. Lassen and northern Sierra areas. On a clear day, the scenery from the top of a snow covered, windswept ridge is beyond description. Glistening snow fields, snow shrouded trees, towering peaks and sky so blue it looks unreal, combine to make unforgettable panoramas of beauty. The windblown snow on snags, rocks and trees form sculptures of all imaginable shapes and forms. It is a photographer's delight.

Most snow courses were established in the 1930's and early 40's, in the headwater drainages of major streams and tributaries. Sites were usually chosen in natural openings in the forest to get accurate snow depths, not affected by trees or terrain. Each course has ten sampling spots, with a tall metal pipe and metal sign at each end of the course. Sampling spots are marked on maps, showing distances and bearings from these markers. Two or three men usually make up the sampling crew. Their working tools consist of 30 inch long sections of hollow and slotted aluminum tubes, graduated into one inch markings. The sections are marked 0-30 inches, 30-60 inches, 60-90 inches on up to 240 inches. Threaded ends allow the sections to be screwed together, until enough tubing is made to measure the depth of the deepest snow expected. A serrated, hardened cutter is at the bottom end of the first section, to enable the tube to cut through ice layers and hardpacked snow. The inside diameter of the tube is approximately 1 ½ inches, calibrated so that one inch of water within the tube would weigh one ounce. The weight of the snow sample within the tube actually indicates the number of inches of water in each sample. Here is how a snow course measurement is done. After reaching the course, the first sampling point is found by measuring a distance from the steel marker to the location of the first sample. The tube is assembled with the inch markers in sequence, an aluminum handle is bolted in place to aid in pushing the tube through the snow, and we're ready to take the first sample. The tube is plunged vertically into the snow until the ground can be felt. The depth of the snow is noted from the calibrations on the outside of the tube. A clockwise rotation is made to "lock" the snow core inside the tube. Then the tube is carefully lifted out of the hole, so as to not lose any of the core. The length of the core is then measured by checking through the slots in the tube. There is always a certain amount of compression of the core within the tube, but an acceptable core length is no less than 90 percent of the snow depth. This length is also recorded. The weight of the full tube is made by carefully balancing the tube on a cradle attached to a scale held by one of the crew. This figure is recorded, and the core emptied from the tube. The weight of the empty tube is then recorded, and subtracted from the weight of the full tube, the difference being the inches of water in the sample, or water content. By dividing the water content by the total depth, a density percentage is obtained, which varies from about 10 percent in new fallen snow to over 50 percent in late spring, solid packed snow. This process is repeated at each of the ten sampling points, with results totaled and averaged to come up with a snow depth, water content in inches, and a density percentage for the course.

Data from approximately 20 courses on the Shasta-Trinity National Forest is taken each month and relayed by telephone to the California Department of Water Resources in Sacramento. From this data fed into computers, information on amounts of water for power, irrigation, recreation and flood control can be calculated and dispensed to the areas, agencies and public utilities for their use in determining their individual needs.

"The Long Trip," it was called, and rightly so—six days and five nights of cross country skiing over some of the most beautiful country imaginable. The five nights spent in different snow survey cabins, and the six days traversing steep ridges, and long slopes, through windswept passes and over blue-white cornices on the downwind sides of sharp ridges. Spectacular views, and an element of danger from avalanches made this a very memorable experience. I made this trip on three occasions. Each time with Tom B., and with a different third person, Don S. in 1960, Chuck C. in 1961, and Fletch in 1962.

Don, Tom and I reached the Parks Creek cabin in a heavy snowfall, having taken the Sweetwater Course enroute to the cabin. It was an easy day of only four miles. The Parks Creek cabin was a deteriorating log structure, dark and gloomy, hard to get into because of the snow load on the roof; plus the spongy, rotting, bottom log had caused the front door to be tightly jammed into the floor. After much pushing we got it open, and then had to trim off the bottom to get it closed. A fire in the stove soon had the cabin warmed, chasing away some of the gloom. We decided to go take the Parks Creek Course, since it was early afternoon. The snow continued to fall, but before dark we were back at the cabin, where we ate some supper and rested. Don had some Pinochle cards, so we played a few hands of cutthroat. As I recall, Don was the winner, Tom second, and I was a distant third by the time we decided to sack out. A few mice kept us entertained until we fell asleep. The warmth aroused them from their nests., and the smell of our cooking probably aroused their hunger. Daylight came and the snow was still falling gently as we took off for the Deadfall Lakes and North Fork courses, about 12 miles away and over two 7,000 foot divides. When we reached the Parks Creek Divide we began to wonder if we had made a mistake in leaving the cabin. A real blizzard was in progress. Winds about 50 miles an hour right into our faces caused ice and snow to build up on our goggles and on our eyebrows. The wind and snow numbed our faces and lips as we crossed the barren ridgetop. Frostbite was possible, but a half mile below the ridge got us into timber, where the winds diminished and we could get a little feeling back into our cheeks and lips. All of us had good warm clothing, so we never really got chilled, except for our faces, and in a little while we were quite warm as we approached the Deadfall Course. Don was and is an expert downhill skier and Tom is a good skier as well as being tough as a boot. I had no trouble keeping up going uphill, but my downhill skills were no more than mediocre, so I had trouble keeping up on the downhill runs through the heavy deep snow. At least, the conditions made a soft place to fall, which I did on numerous occasions. The climb over North Fork Divide was not so long as the first climb had been, but fatigue was beginning to tell on us as we neared the top, which we reached without mishap. Tom and Don took off from the top in long graceful turns through the deep snow toward the cabin. I followed at a slower pace, but fell down making a turn and lost a ski. Evidently I hadn't hooked my run-away strap securely and down the hill went my ski, bouncing and bounding through the trees. It was easy to track through the new snow, but walking wasn't easy on one ski, even downhill. About thirty minutes later I reached the ski and got myself together. Another twenty minutes and I was at the cabin. Tom and Don had built a fire and had already taken the North Fork Course when I arrived, so I cooked supper. The North Fork cabin is also a log structure, with a tower on the roof for access into the cabin when snows are deep. It was in much better condition than Parks Creek had been the night before, and didn't have mice. It had pack rats! Big as a cat, they were, and glared at us with red eyes when we shone our flashlights at them. Tom threw a ski boot at one, but missed. Nevertheless, after our pinochle game, we slept soundly. Don was still high, Tom second, and I hadn't gained a bit in the scoring which we decided to continue as a tournament the next three nights.

All of us had blisters on our heels by now and did some first-aid with moleskin adhesive.

Morning came bright and clear, the new deep snow sparkled in the sun and bent the young trees over 'til they touched the snow surface. We felt good as we started up the slope to the divide, heading for Mumbo Basin. This was to be a fairly short day, only about 8 miles and lots of downhill. Snow conditions were pretty bad though, with soft snow and breakable crust as we neared the Mumbo cabin. Scenery was breathtaking, the Trinity Alps stood out sharply about 30 miles distant. We felt like we were the only people in the world, with no sign of civilization as far as we could see in any direction.

The Mumbo cabin was luxurious by comparison. Clean, neat and orderly, with no mice or pack rats. Bud, Henry, and Mel had left a bottle of whiskey with a note in the food barrel for the snow surveyors. That night I gained a few points in the pinochle tournament. I think the "bottle" helped my scoring, as Tom and Don partook a bit of the Jim Beam. Still, I was in last place, but gaining.

The fourth day was beautiful, warm and clear. We made it into the P.G. & E. cabin in early afternoon, built a fire and took the Whalan Course while the cabin warmed up. Again we played pinochle, with a different deck, and one of the cards had "Martha" written on it. The scores evened up and we were nip"n"tuck by bedtime, with Tom in a slight lead. We didn't sleep as well this night. The cabin was tall, with a loft, the bottom floor was too cold, and the loft too hot, but we survived.

On to Highland Lakes the next day. Uneventful, and we reached the cabin after lots of ups and downs over the ridges, taking the Highland Course on the way. We were getting tougher as we went, but the miles were telling on us and the sun and wind had weathered our faces to a leathery texture.

The final night of the tournament Don had slipped to last place, and Tom and I were battling out for the winner prize, which was to be "Martha's" cards. We predetermined a number of hands to play that night, and when the last hand was dealt it was obvious that whoever got the bid was going to win. Don quit bidding as Tom and I kept going. We each had a good hand with a run and the bidding got up into the high forties. I really hadn't seen the 100 aces in my hand when Tom quit bidding. He figured he could set me, and I figured I was going to have to pick 23 points as I put my meld down. Then I saw the fourth ace in my hand, which cinched me making my bid. "Hey, I have a hundred aces," I exclaimed, as I put my entire hand down. Tom looked in disbelief. He made some unseemly remarks as he grabbed all the cards and flung them into the rafters and walked out the front door. The six or so feet of snow outside soon cooled him off, as he sheepishly came in the cabin. Now, Tom always did have a hair trigger temper, but was equally quick to forgive, so before the night was over, even Tom was looking for the scattered cards. All but one was soon recovered, and we searched and searched to no avail. Finally after giving up and sitting around the stove for a while, one of us glanced up, and there was that card, sticking in a crack in the rafter.

Yesterday's thaw, and the freezing night had made the snow hard as a rock when we left for home the next morning. Barreling along at a rapid pace on the frozen snow just after leaving the cabin, Tom hooked a tip and took a hard spill. He twisted his ankle very badly and we were miles from our transportation. Barely able to put any weight on the ankle, Don and I helped support him for a long time. Finally he was able to hobble slowly on the injured leg, but it still took us hours to reach our pickup. Though there were no broken bones, Tom limped for several months after the accident.

M-7 Snow Tractor, 1950

SNOW VEHICLES

An M-7, it was called, the first practical oversnow vehicle on the Shasta Forest. The army had developed it as a winter vehicle for the invasion of Norway by the Allies, to be dropped from airplanes with paratroopers. As I look back, remembering its design, it should rank as a "classic", somewhat like the vintage automobiles which have that certain something distinguishing them from the run-of-the-mill types. It was long, lean, and a convertible to boot. There were two skis on the front, controlled by a steering post and wheel at the driver's seat to assist in turning. Rubber coated metal tracks with a power sprocket were the drivers. Two single in-line seats crowded between the tracks made up the driver and passenger compartments. A tubular frame with a canvas roof and twist snaps to attach the isinglass windows, kept the cold air out. These windows could be rolled up or removed as the temperature dictated. Extra passengers could sit on wooden fenders over the tracks, and hang onto the tubular frame, but most times, extra passengers were towed behind on skis with a long rope. When snow was good, this was a going machine, enabling former two day ski trips to be completed in a matter of hours. But, deep soft snow tended to bog it down, and the narrow gauge tracks contributed to lots of side slipping, and a threat of rolling over. Much shovel work was needed to make a rut for the upper track, as we maneuvered around steep side slopes. Oftentimes it was easier to abandon the snow tractor and go ahead on foot, but as most humans are, we would rather work for a couple of hours in order to use a machine than to ski uphill for an hour.

Late one warm winter day, Bob M. and I took the snow course measurements at Sand Flat and Horse Camp. Snow was melting at a rapid rate, with water literally pouring out of the snow and running underneath the snowpack. By the time we got back to Sand Flat from Horse Camp, the snow was like slush. As we traveled in the M-7, wet snow and water was flying from the tracks. Suddenly, the bottom went out in the soggy snow and we settled into about two feet of water and ice, about the texture of soft mush. There must have been enough ice under the vehicle to high center it, for the tracks just churned in the slush. I was driving. "Now you've done it," says Bob, as the water rose under our feet and we eased a little higher in the seat and onto the running boards. I can't remember for sure, but I think the water got over our boots as we scrambled over to some dry land about 40 feet away, where the snow had melted around a large tree. We contemplated our predicament a moment until we spied some broken limbs and chunks of wood under the tree. With little hope of getting out, we waded back to the vehicle with all the wood we could carry and started poking it under the tracks. Wet, cold, and with little confidence, we started the engine, put it in gear and drove right out of our mess. If I recall correctly, Bob drove the rest of the way and I didn't protest. We remained friends.

A "committee" must have been designers of the Sno-Ball. No one person could have come up with such a machine. Powered by a high powered Chevy V-8 engine, its other components were the cab of a Jeep, the transmission of a Dodge, the rear end axle, drive and housing of an Oliver Farm Tractor, and a few Ford parts, plus lots of Rube Goldberg cables and links designed or stolen from other over-snow tractors. The Sno-Ball was a pretty unpredictable rig. The only certainty as a trip began, was that it would break down somewhere along the way and never twice in the same manner. Tracks would roll off the sprocket, steering cables would come off their pulleys, the radiator would get too hot, a spring would break, or, on one occasion, the rear end housing of cast iron crumbled under the strain of too much power from the engine. Other than these difficulties with extra wide tracks, it was probably the fastest and most maneuverable snow tractor we ever used, but we only kept it one winter because of its tempermental nature.

Matt, the mechanic from the Division of Water Resources loved the machine. He liked the monthly trips to Mt. Shasta to bring the Sno-Ball in.

Old Reliable, the Thiokol ten pasenger snow tractor seldom failed us, and since 1965 has been the mainstay in the snow survey program on the Sacramento and McCloud ranger districts. A boxy looking machine, powered by a Falcon industrial six-cylinder engine, it is so well designed and adequately powered that little mechanical trouble has been encountered, but a few operator caused mishaps have occurred.

When a machine is new, caution and lack of knowledge of its capabilities tend to keep a man out of trouble, but as the machine becomes more familiar, the more daring one becomes in trying the impossible. Snow conditions make big differences in effective use of any snow tractor. With ideal snow, impossible looking maneuvers are easily accomplished, and conversely, poor snow makes apparently easy looking travel impossible. The driver soon gets the feel of the machine and the snows, but some get it quicker than others.

Sidehill travel on packed snow can be a thrill, or downright scary, with a deep canyon below you, and the snow cat tracks side-slipping across the slope. Most travel to snow courses is on roads, but wind currents cause drifting snow to fill in road cuts to the point the road is not discernable, and the side slopes often exceed 50 percent. For short distances, under good snow conditions, the operator learns to keep the nose of the snow cat headed uphill as he pours the coal to the accelerator, thus making the machine slither across the slopes like a sidewinder, with snow flying and tracks spinning. When a slope like this is encountered, the passengers usually get out and let the operator take it across by himself. The Deadfall, Parks Creek road has one of these scary places, where, if you lost momentum, down the machine would slide for several hundred feet with nothing to stop the slipping. Ben and Tom and I approached this tricky area one day. Tom was driving. "Sure, you can make it," I tried to persuade him, as I plumbed the slope with a shovel handle from outside the machine. "It's only about 50 percent and if you keep her head up, you'll make it O.K."

"Do you really think so?"

"Sure, I do."

"All right then, she's all yours. Take it across," says Tom as he climbed out from the driver's seat. Ben grinned from a safe distance and didn't say a word. Well, we, or should I say, I, made it. We decided to take a coffee break after the successful crossing, and to contemplate the return trip later in the day. The remainder of the trip was uneventful, as by the time we got back, the snow had softened to the point the machine had good traction as Tom drove across. Now don't get me wrong, Tom is probably the most competent snow cat operator we have, and he delighted in talking me into doing the job he could have probably have done better.

The one time the Thiokol really let us down, it did so in grand fashion. The Division of Water Resources people decided to make a T.V. documentary film, photographing and narrating as we took the Sweetwater, Parks Creek, and Deadfall courses. We really cleaned up and loaded down the snow cat with everything from TV cameras, boxes of film, several cameramen, a guest from the California Division of Forestry, three wheels from the Division of Water Resources, plus Ben, Tom and me to do the work. Things were going great; the weather was beautiful, the scenery unbelievable and we were making pretty good time. Suddenly there was a strange grinding noise in the rear end, our vehicle came to a screeching halt. By feeling the final drive housing, we knew something was badly wrong. It was so hot the paint was smoking, and a handful of snow thrown on it would melt in seconds.

To put it mildly, we were in a predicament. The day was fleeting by, the courses not taken, and here were nine men stranded some 16 miles from the trucks. After a huddle and some discussion, we decided that five of the nine who were in good shape would head for home on skis, since there were only about three miles of uphill travel, and thirteen miles of downhill in well beaten snow cat tracks. I stayed with Jean B. (a man) and two others from the Division of Water Resources. We planned to sleep in the snow cat and go out the next morning on foot, because the three of them agreed they could cover the sixteen miles if they had plenty of time to do it.

Before dark, the four of us skied about a mile to take the Deadfall Course, and returned to the broken down snow cat. By then it was getting late afternoon, and cooling off rapidly, so we dragged down some dead wood from trees above the rig, and built a warming fire just outside the snow cat. There were a couple of blankets under the seats, plus a few Forest Service rations for just such emergencies as this, so we were in good spirits as we rearranged the seats and cushions to make beds for three. I volunteered to sleep in the driver's seat, and to keep the vehicle warm by running the engine and heater occasionally during the night. Since we had plenty of fuel, and nothing was wrong with the motor, everything was ready for the long night.

To pass the time, we decided to play a little poker before going to bed. After the game, the three of them decided that with my winning all the money after having to borrow a dollar from one of them to get started was just too much, on top of stranding them out in the boonies.

The night passed uneventfully and everyone slept well except me. The front seat was sort of cramped for my long legs, and having to start the engine every hour or so kept me awake. By morning I was stiff and sore, while the rest of them were rested and relaxed. "Serves you right, for winning all the money," was their sarcastic comment.

Just about daylight a message came over the portable radio that they were sending a copter to pick us up. The press had been told that "Bob Gray and a bunch of wheels were stranded in the mountains in deep snow without food," plus all kinds of other newsy details. Humiliating!! That's what it was. When we reached the truck, there was Alva, the news reporter, grinning at me while shooting away with her T.V. camera.

All ended well. Everyone got home safely, we got the courses taken, and I survived the embarrassment of the T.V. report. We towed the broken down Thiokol out with a Tucker snow cat brought up from Sacramento.

Just another example of the catastrophies that occur on snow surveys when big shots go along.

SNOW SURVEYING

The middle 40's were still the non-mechanized days of the Forest Service, and all snow surveys were taken on foot. From Highway 89 to Stout's Meadow was a tough 12 mile uphill grind on skis. My first couple of trips were with Cactus. I soon became quite proficient at cross country skiing with a pack on my back, so I really liked the monthly trips into Stouts Meadow. We had built a comfortable cabin there in the fall of 1945 and staying a night in a nice warm cabin in the muffled quietness of the snow covered shelter was quite pleasant. After leaving the highway at about six a.m. we would arrive at the meadow about two p.m. if the conditions were good, but sometimes after dark if the snow was soft and sticky. Snow was very deep my first trip up, with heavy snow loads bending trees over to the middle of the narrow road. We stopped to rest at one point, and just to my left was a deep hole alongside a large tree trunk. Snow had blown and drifted, creating the hole which was about eight feet deep and three feet wide. "Cactus," I said, "Look at that hole, I'd hate to fall in there." No sooner had the words left my lips when the ski pole I was leaning on gave way. I tumbled into that hole headfirst, pack and all. My skis lay across the top with me hanging head down. With the pack dangling around my arms, I was helpless, as I couldn't shake my feet from the beartrap bindings. Cactus heard my muffled shouts, and peered into the dark hole. His humor differed from mine about the situation, but he did pull me out after gazing in wonder at how I'd managed to get in that predicament.

The rest of the trip up was uneventful, but I did get a nice photo of Cactus waxing his skis at another resting spot. About one foot of a 16 foot telephone pole was sticking through the snow, and the single strand of wire was at snow level. It was the deepest snow I'd ever seen at the time.

Now, coming down from there was something else. The tracks we made in fresh snow going up froze solid during the night. Those skis went unbelievably fast downhill in the frozen tracks, much faster than my skiing capabilities had developed. The road was so narrow from leaning trees we couldn't get out of the tracks from the day before. Even Cactus was having trouble. Our solution was to cut two poles, and ride them like a stickhorse between our legs and skis. With this we managed several miles of steep fast going. By sitting on the stick, the weight on the dragging end would keep our speed under control.

Another trip to Stout's Meadow with Jack almost resulted in a disaster. The snow conditions were too good. A light skift of new snow covered an old hard base and the road was like greased lightning. Jack was a better than average skier for those days, and he was really turning them loose down the hill, but even for Jack there was just not enough maneuvering room on the narrow road. He caught a ski tip on something and took a header into an exposed rock. His head and shoulders hit the rock and stunned him for several moments. He was really hurting by the time we found his glasses and got his gear together. I carried both our packs the eight or nine miles to the road, because Jack had severely strained the muscles in his shoulder and upper arm. Jack couldn't ski, so we had to walk. Fortunately the snow was hardpacked, making walking fairly easy.

George and I took a snow survey together, and it too turned out to be less than pleasant. The early morning was beautiful, but about three hours out a real blizzard hit us, limiting visibility to about 20 feet, but it was a course I'd taken several times and I was sure I knew the way. We continued and so did the storm. About two feet of new snow piled up and the going was tough. About four p.m. I thought we should be reaching the cabin, because we had already passed a landmark I was familiar with and it was only about a mile from the cabin. Suddenly in the growing darkness, I saw where someone had passed right in front of us on skis. George was too tired to notice, and then I realized those were our own tracks. We had made a complete circle, and the snow had piled up in the tracks so it was difficult to tell which way they were going. Fortunately I recognized another landmark, and we got headed in the right direction again. About 100 yards from there I could see where I made the mistake in making the circle. Thirty minutes later and in total darkness, we arrived at the cabin, utterly beat, but elated that we didn't have to spend the night in the snow.

We slept late the next morning and took the course about ten o'clock. Our troubles started again, the tube broke at the first joint about eight feet down in the snow. It took us about two hours to retrieve the cutter section and finish the course. When we left the cabin about noon we were already tired, and the new snow made going downhill hard. Our tracks were totally obliterated, and by the time we reached our Jeep, it was dark, *and the battery was dead.*

It was still two miles to the highway, which took us another hour to travel. Just as we reached the highway a car appeared. He passed us by like a dirty shirt. That was the only traffic until we had walked another three miles toward the Mt. Shasta Nursery on Highway 89. A search party was just being formed for us as we knocked on Karl's door after getting a ride for one mile. Our wives were pretty upset by this time, but relieved to hear our voices over the phone. The next morning we took a battery in to our stranded Jeep. George had enough of snow surveying for the rest of his Forest Service career.

Resourcefulness is a trait born of necessity in Forest Service employees. Lack of funds, lack of materials, and lack of the proper tool could always be overcome by such persons as Paul and Andy. It seems they arrived at the snow survey cabin expecting to find the snow sampling equipment in its usual place, but no, they had taken it down the previous month to do some repairs and forgot about it. Well, Paul and Andy weren't about to let a minor detail like that thwart their efforts. And another thing, if Chuck, their boss, a less than understanding individual, knew that four man days were wasted by a dumb human error like forgetting, he would never forget it.

A few moments of contemplation, and they hit upon a plan. It was easy to get snow depths by using an iron rod, and checking the mark on the rod by comparing it with something of known length, like for instance, the length of Paul's boot, or the width of several one inch by eight inch boards. To get such things as water content would be something else, though. They took a section of six inch stovepipe, and pushed it down through the snow, carefully removing the snow and putting it into some containers. This was repeated until all the snow had been removed from the six inch diameter hole from the snow surface to the ground. By melting the snow in all the containers they accumulated all the water from a cylinder six inches by "x" inches. Being a couple of pretty smart non-professional foresters, they were able to calculate the volume of the hole, and the volume of the liquid in the containers. By dividing the volume of the hole into the volume of the "distillate" they came up with a density percentage figure, which to their trained eye seemed quite reasonable. Using this percentage against the calculated depth of the hole in inches, they came up with a "Water Content in Inches" figure. Admittedly, this one hole might not have been 100 percent representative of the course, but it was close enough for this one time, and anyway, next month the course was to be measured again, which would be somewhat of a check on their improvised sample this month. The next month's measurements bore out the fact that they were pretty much "in the ball park" with their unique calculations. Chuck, you should be proud of your men, "They done a good job!"

Seeing wild animals while on snow surveys is really quite rare, but occasionally some hardy creature is seen, or at least his presence is noted. A pine martin, peering from a limb about 20 feet above the snow is a rare sight, but I have seen two or three, and their tracks in the snow are fairly common. Snowshoe rabbits, with their white winter coats are more common, and I've had them run right across my ski tips while sliding downhill—a quite startling sight. Ben and Tom and I decided to chase one on skis near Gray Rocks Lake, after he had passed Ben like a streak. Needless to say, we lost him in a fir thicket, but the chase was fun. It was the first one Ben had ever seen, and he showed up just minutes after we had been talking about the color-changing creatures.

In the late winter or early spring season, a coyote may be seen loping over the hard-packed snow after one of the snowshoes, or after a pine squirrel who has had enough of hibernation. And, the dumb old grouse will hang out in his fir thicket all winter long, not knowing that good weather is just down the hill a few miles.

I never saw a bear on a snow survey, but one made his presence known on a snow survey trip to Stout's Meadow. The Thiokol snow cat was new, but had broken down, so the crew skied to the highway to get some tools to fix the machine. It was the next day when they arrived back at the scene. Somehow, a bear had broken into the snow tractor and had utterly destroyed the upholstery and foam rubber pads on the seats. The naugahyde was ripped to shreads and foam rubber was in little pieces. In addition, he had broken two windows and had eaten all the rations in the machine. After the crew left, because they had to return the third day for more repairs, the bear went through his act again, but there wasn't much left to destroy the second time around, except to clean out a few more rations.

One of the most unexpected bit of wildlife encountered on a snow survey was while going to Stout's Meadow in the old M-7 snow cat. Snow was falling heavily when we reached the Tate Creek crossing in a dense, tall, fir timber stand. Here, standing and walking around in the deep snow was a flock of wild geese—Canadian honkers, if my memory serves me well. They paid little attention to us and were still there when we came back by several hours later. They must have been forced down in the late winter blizzard.

SAND BLASTING

After the fall of 1947, whenever I inherited a terrible Forest Service chore, I could look back and truthfully say, "Well, it ain't as bad as sandblasting."

Ted, Jerry, Ray and I were given a good winter job in Redding. No time off, and plenty to keep us busy. It sounded great, because few Forest Service temporary employees ever got year long employment in those days. We had most of the necessary equipment, a large air compressor, a sand tank, hoses, nozzles, respirators, goggles and a mountain of sand bags. This, plus hundreds of gallons of red lead paint made up our project for the winter.

Three men rotate the jobs of nozzleman, steel man, and sand man. The sand man keeps the hopper to the sand tank full, the steel man operates the cherry picker to place the rusty bridge steel in place for blasting, and the nozzleman blasts the peeling paint and rust from the steel. Our instructions were "to sandblast and red lead all the bridge steel in the yard at the Forest Service shops, and this is what you do it with." No one showed us how to do it, much less, how to do it safely. We learned by doing. Efficiency increased as time went on. Fortunately, no one was injured seriously. The job was hard physically, and dangerous, as an accidental blast of sand from the nozzle could cause serious injury. Balancing and painting them was hazardous, too. An extra hand would have helped.

The worst feature of the job was that, even after a shower, I would wake up with sand in my bed. On well, we survived the winter with nothing more serious than a few trips to the doctor to remove sand from eyes, and we even got a nice letter of commendation from the engineering department of the regional office for the good job. One good thing about the job was making new, lasting friendships with men in the Forest Service shops and fire cache near our operation.

BEAR PROOFING SIGNS

As late as the mid-fifties, bears were still our number one vandals in the woods, with people coming on fast. Paul replaced Cactus at McCloud in 1948, and took over fire control, which at that time included most everything. Our sign program had deteriorated badly under Dutch and Cactus, not because of inefficiency, but those men just didn't need signs. "You turn off the road where the coyote crossing is," likely would be their direction to a fire. Both had been there so long that a lack of directional signing seemed rather unimportant.

Bear had done more than their share of raising havoc with the signs. They don't eat them, but rip them to shreds with their teeth just for the fun of it, and scatter the pieces from hither to yon. Paul had an idea, and it worked. We cut thousands of four inch pieces of no. 12 rigid copperclad telephone wires, which we drove at various angles into the edges of signs, until they looked like a wooden porcupine with writing on them. Some of these signs put up in 1949 are still in place.

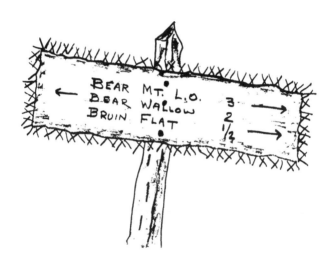

A Bear-Proofed Sign

COUNTING SHEEP

Not all sheep counting is done while trying to go to sleep. The national forests of the West are often summer range for both cattle and sheep, which spend their winters in the lowlands of the great valleys, or hanging in the slaughter houses. Mountain meadows are lush with nourishing grasses and browse, which are highly prized by stockmen and ranchers. Those fortunate enough to obtain grazing allotments from the Forest Service have low cost, quality feed for about four months each year, based on a few cents per animal per month. Spring snowpack and range conditions checked during the summer determine the time livestock go onto the range and how long they stay. Overgrazing seriously affects the fragile meadows, and the stock has to be kept moving so as not to remain in one location too long. Some of the allotments are entirely accessible by road, while others are in more or less roadless areas, and require the services of a line rider to keep track of and move the stock. One such man is Si, a very interesting man who lives in a primitive log cabin with his horses and dogs during the summer grazing season. Most of the time he is by himself, but on occasion he is visited by his wife, or daughter, or Ethel, a very outstanding lady who owns the cattle. More about her later.

During a busman's holiday last summer, 1976, I took my wife, daughter-in-law, daughter, son-in-law and grandson to see Si. We had gotten wet and cold taking a hike in an unusual August rain, when I thought about going to his cabin. The warmth from the big iron cookstove, and the sound of rain on the roof, plus Si's warm hospitality made the day worthwhile. Except for the knowledge of the stock truck and car up on the road, it was like stepping back in time 100 years. The memory of this family outing will always be in my heart and mind. A picturesque cabin with smoke coming out the chimney, a pole corral, and a barn sagging in various places, the smell of wet horses, two alert cow dogs and the little leaning outhouse all added to the occasion. We hated to leave.

Ethel, the owner of the cattle, and the holder of the grazing allotment, is a very gracious lady who is also a packer, guide and cattlewoman of renown, and at the same time a remarkable and respected woman. She and her husband live in Red Bluff, but she rides her allotment quite regularly, fair weather or foul.

Sheep have all but disappeared on the Shasta-Trinity, but in the 40's, sheep still browsed through the McCloud flats, with a dust haze continually hovering over them as they moved in large herds.

Dutch sent me to Bartle one spring day to count sheep as they were unloaded from the McCloud train. I stood at the bottom of the unloading chute as they were coming out of the rail cars, counting as they crowded by. A tally whacker is a counting device held in your hand with a push button to record each animal, and the total shows on the dial. After tallying several car loads, I checked the dial. I had tallied about twice as many as the sheep allotment allowed. This shook me up, and I told the owner he had far too many sheep for the allotment. "No, in fact I'm a few shy," he said, then a puzzled look came on his face. "Did you count the lambs, too?" Well, no one had told me to only count the ewes, so I had counted everything. Anyway the lambs were as big as the ewes, because the ewes had been shorn and the lambs were still "wooly." Dutch laughed at our mistake, and we accepted the rancher's figures. There was no fee for the lambs, so they were not to be counted. Now I know how to count sheep.

PREVENTION & CONSERVATION PROGRAMS

Most fire prevention and conservation programs are aimed at young people. They're more receptive and much less cynical than even the high school age groups. Prevention and conservation are year-long jobs, but during the springtime, before school is out, we try to schedule programs with all the elementary schools on the ranger district. Fire prevention is stressed in ways acceptable to the age groups. Smokey Bear coloring books, and song sheets appeal to some; movies and animated cartoons appeal to others. All ages like to actively participate in the programs, so question and answer periods are allowed, usually with questions written on paper to cut down everyone talking at once and to assure everyone getting an answer. At older grades, and even in high school and adult groups, panel discussions are possible, with several specialists from the Forest Service who can answer questions about recreation, timber, fire, mining, wildlife or whatever.

To add to interest, sometimes we take a smoke-jumper's suit, parachute, and pack to show that part of fire control. Maybe a cut from a tree is shown to illustrate how a tree grows. An assortment of cones or needles will help a group identify the variety of conifers in a certain area. Even a personal experience true story can add to the program.

Some of the most rewarding, and best received programs were those at the Methodist Camp on Scott Camp Creek. Family groups attended, usually in the evenings around a bright campfire, or in the barnlike lodge in front of a stone fireplace. Singing and refreshments usually followed these programs, and I always enjoyed participating. Conservation and nature talks were the big thing at these meetings, with a little fire prevention subtly included. One minister would always sing a song about "The Forest Ranger" when I arrived; to make me feel at home, I guess. Not all of these programs had the desired results in fire prevention. On a couple of occasions, we had fires within a week following the program—started by a boy or girl within the group.

It was in the early sixties that I was selected to be one of a committee of four judges in a fire prevention poster contest at the Dunsmuir Elementary School. Now this contest was part of a Forest Service Conservation Week program which I had presented. Our committee consisted of Nick, the fire chief, the chamber of commerce president, the local art teacher and me. Soon we had the posters narrowed down to the final few. It happened that one of the finalists was my son, David, of the kindergarden class. "Oh, come on," I said, "we can't have my young'uns poster in the finals."

Well, I argued to no avail. His was selected over my protests, and the local weekly newspaper didn't help my uncomfortable position when it listed the poster winners, and the judges names. I was pretty proud of his poster, and picture in the paper, though. He deserved to win.

RECONNAISSANCE FLIGHTS

Most people enjoy flying, and recon flights are fun. Their purpose is to find something. In my fire control job, the primary purpose has been to find lightning fires outside the visible areas of forest lookouts, or to determine the potential of fires as they occur. Recons are taken for other purposes, too. Looking for lost persons, checking for outbreaks of insects or disease in trees, or even checking snow conditions in the high country. In 1962, after the Columbus Day Storm, flights were made to determine areas and volume of blown-down timber, which amounted to millions of board feet over the forests.

With an estimated 150 recons to look back on, there were some memorable trips. Most have been in contract ships, but a few in Forest Service planes with Forest Service pilots.

After a thunderstorm, the lookouts inform you of the general path of the storm, and the location of any observed downstrikes of suspected smokes. Any fires reported by lookouts are usually checked first to ascertain their locations, unless crews are already on the fire.

Otherwise, a pre-arranged flight pattern is flown for 100 percent coverage of the district. Many of my first recons were in a Cessna 195, with Alma, a lady pilot and a good one. She had flown cargo planes to Europe during World War II. She flew conservatively, at fairly high altitude, unless we were marking fires with toilet paper, or had to get down to see conditions at close range. Alma told me about eagles and airplanes. She had either hit one, or had a near miss when trying to avoid one, thus learning that you let the eagle avoid the plane and there will be no trouble. Her flying technique was such that the observer could see every area as she smoothly flew a wavy pattern along the route, leaving no blind areas in front or behind.

Gordon was another Cessna 195 pilot. He could quickly remove seats and a door from his ship to convert to a small cargo dropping plane, which we did on several occasions. We had discovered a fire in the lava bed country, and by the time a crew reached it about an acre was burning. The crew was going to be there a day or two and would need more supplies than they had carried in, so Gordon and I converted his ship to a cargo transport, and I was to be the dropper. Pushing the small packages out the door was easy, as he had static lines to open the chutes on the cargo, and a webbed belt with a snap hook to tie into my harness so I wouldn't go out with the cargo. We made about a dozen drops of five gallon water cans, food, and equipment from a height of about 200 feet. As the packages were pushed out, I'd stick my head and shoulders out the door to see where the chutes were landing. The cabin was not large so I was on my knees with my hands on the floor at the threshold after each package was released. When the last cargo was gone, I sat back on my knees and feet. There, hanging in front of my face, was the webbed belt and hook, which I thought I had been attached to all that time. Gordon and I both saw it simultaneously. I don't know yet whether we had ever hooked it, or if it became unhooked during the drops. He was more shook-up than I was.

At Weaverville, I never tired of recons. It was some of the most spectacular country imaginable. Lakes, all sizes and shapes. Some were like emeralds, some blue, some green, and all beautiful. The Trinity Alps wilderness area had everything. Craggy peaks, rockslides, deep canyons, turbulent streams, lush meadows and trees all were there. Wild game abounded. In early summer, velvet antlered deer could be seen in most meadows, and even bear making trails through the tall wet grass.

Lonnie knew how to use the thermals in the steep rock canyons to help lift his plane over the ridges and into the next drainage. I learned the whole Trinity side of the forest from the air, as the neighboring districts would ask us to take a look at remote areas for sleeper lightning fires. This is real smoke-jumper country, where men travel by plane and brightly colored chutes into the fires. On one occasion with Lonnie, the thermals just weren't there, and we had a tight squeeze getting turned around in the box end of the Stewart's Fork. It would have been a beautiful place to end it all, but we didn't relish the idea and managed to get turned around and headed down the canyon. Lonnie is a skilled pilot, but we flew a little higher after that.

Another incident happened with Lonnie about 16 years later. Two of my friends, Bob M. and Bob B. were on recon over the Mt. Shasta area when the throttle stuck at full speed. A pin in the linkage fell out or broke, and Lonnie couldn't slow the motor. This altered the flight plan considerably as they headed for Redding. Bob and Bob later said they weren't too shook up, as they had confidence in Lonnie, who assured them he could handle a dead stick landing at Redding. Nevertheless, we arranged for emergency equipment at Redding in case things didn't go as planned. All these arrangements didn't go unnoticed as Bob B.'s wife heard the plan over her monitor radio set. Naturally, she was upset and it took some persuasion and convincing by me that there was really no danger, and the boys would land safely. They did. Loretta later told me I didn't sound too convincing on the phone.

TALES ABOUT A COW, A CALF AND A WEASEL

Rangers get called on to do everything. Like rescuing a new calf in the deep snow. "Just call the Forest Service, they'll do it," and they usually do.

It was late fall, and the three-foot snow had covered the McCloud area when Paul, the fire control assistant, got word of a cow and her newborn calf floundering around in the snow about four or five miles out of McCloud. No, this "weasel" wasn't an animal, it was an army surplus, open air, tracklaying vehicle which the Forest Service had acquired. Paul and I took off from the ranger station headed toward Mud Creek in the weasel. Sure enough we found a furrow in the deep soft snow, obviously made by a cow and calf. We followed the tracks in erratic circles until we found the mama and her baby under a pine tree. If there was anything that cow didn't want, it was anything to do with us. We figured on herding her and the calf back to McCloud in the tracks we'd made coming out. Cooperate? Not that old heifer! She glared at us, shook her horns at us, and refused to do anything but circle those trees, with us in hot pursuit in the weasel. The calf watched in curious amazement. What kind of a world have I been born into? he must have been thinking.

After failing miserably to herd that old gal in any direction, I told Paul to let me out and I'd lasso the cow, then we could lead her home and the calf would follow. Well, I ran at her with my lariat swinging over my head. No cow is supposed to react as she did when the noose settled around her horns. Instead of pulling away from me, she charged. Now, we were in three feet of snow, you know, and I wasn't prepared for a charge. As I back pedaled in my tracks she came on like a locomotive, hit me in the chest with her head, knocking me backwards about five feet on my back. Fortunately for me, she also stumbled and fell, giving me a chance to scramble to my feet. What was Paul doing all this time? He was laughing so hard that tears were rolling down his cheeks! I think he did manage to ask me if I was hurt before continuing with his uncontrolled laughter. When I finished brushing the snow out of my collar, I climbed into the weasel, still holding onto the rope. Since I had no injuries, no thanks to Paul, the situation began to appear funny to me, too.

Old Bossy wasn't about to be led by that weasel. I guess she thought we were going to take her away from her baby, so she laid down. Well, we thought, if we tie the rope to the hitch she'll soon get up and walk. She didn't. After the rope stretched, her neck started stretching too. Did you know that a cow's neck can stretch about twice its length without any bad effects? We dragged her through the snow, but she would not get up. The calf stayed under the tree. We simultaneously thought, if we load the calf in the weasel, that old biddy will follow us. The calf sat in the back seat, and we finally got the cow on her feet. An hour or two later we got to the barn in McCloud with our livestock. All of us totally tuckered out except the calf who was probably beginning to think that this is a pretty interesting world.

The rope was still around the horns of the cow when we released her into the barn, and it was too dark to try to get it off then. I'll get it off in the morning, I thought, and I did, but not before she had her last fling at getting even for the injustices of the day before. When I went into the barn I took the end of the rope and snubbed her to a post so I could get the rope off without her getting me. She pushed her head against my ring finger and mashed my ring so that I couldn't get it off. I finally succeeded in getting it off by straightening it with a pair of pliers.

The owner of the cow and calf came and picked them up without a thank you, or any comment.

AN ERRAND OF MERCY

"An exclusive, fisherman's paradise, owned by a bunch of millionaires," was the way most of McCloud referred to the McCloud River Club. It was, long before our "ski-in" in 1957, and still is, a fabulous, almost mysterious landmark on the lower McCloud River. Fabulous, because of the primitive beauty of the river and surrounding country, and mysterious, because of the unavailability to all but a selected few. Sometime after the turn of the century, rumor tells us, an exclusive group of a dozen men acquired the land, and got permission from the U.S. government to build a private road from the Shasta County line south of McCloud to the McCloud River Club. This must have been a tremendous undertaking at the time, as the terrain is steep, rocky, and covered with dense forests. Approximately twelve bridge structures cross Squaw Valley Creek and various tributaries before reaching the McCloud, where another large bridge crosses the McCloud itself. Even in summer, the driving time to the club over this narrow, winding road of about 20 miles took several hours.

Locked gates at the Siskiyou-Shasta county lines keep tourists and sightseers out, and even though the road goes through several miles of national forest land, the public is excluded. A foot trail, also through posted private lands, leads from the McCloud arm at Shasta Lake, up river to the club, some 15 miles distant.

Heavy, wet snows fell in the winter of '52, and several persons were isolated at the club, with no communication and a rumor that there was illness among them. One of the owners in the Bay Area contacted Mr. G., a wheel in the McCloud River Lumber Company, asking if he could find anyone to ski to the club, to take mail, medicine, and to find out the condition of the sick.

Harry, a tough, wiry ex-merchant marine; Jack, one of Dutch's sons, and I volunteered to make the trip, each taking leave of our normal duties to make a few extra dollars.

The wet snow falling as we started skiing in the early morning rapidly turned to rain, so we put rain clothes on over our outside clothes. Before long we were wet from perspiration inside, and rain outside, but remained warm except when we stopped to rest. So we kept going. Skiing was terrible, the wet snow clung to our skis and had also bent the limber trees over into the road, blocking our way. We either had to climb over them, or knock the snow off so they would rise back up out of the way. Either effort took its toll on our strength and endurance, and we had to continually change places to break trail, thus having one working and the other two resting. The resting wasn't really easy, just easier than breaking trail.

It was late afternoon when we reached our destination, tired and wet. The marooned people were surprised to see us and had pretty much recuperated from their ills, but were glad to get mail and the medicines.

Next morning dawned bright and sunny, so with dry clothes and boots and a full stomach we headed home. Going back was more pleasant, as the trail was made and the weather much nicer. Uphill made a difference though, and we reached Braden cabin in late afternoon, with several miles to go. The cabin, which we had bypassed on the way down, was too tempting to pass up, so we spent a luxurious night before a cheery fire and in warm beds. A three hour trip the next morning took us to McCloud, where we reported the conditions of the snow and the people at the club. Mr. G. was not in the office, so we talked to Elmer, the woodsboss of McCloud Company woods operations. He decided that $100 each was not too much for our three days and that those millionaires could afford it. Therefore we received more than expected for our labors.

About two or three weeks later, one of the club owners, thinking our cost had been too high, hired a man to make the same trip alone. He was not familiar with the road, and had never been to the club. When he failed to appear back in McCloud after four days, the Forest Service was asked to take the M-7 snow cat down to see what happened to the man. His body was found by Paul and me in the middle of the road, one mile from the destination. We think he lost confidence in where he was going, not realizing he was almost there, and probably died from overexertion and fear.

THE MT. SHASTA NURSERY

"We're going to clear out all these trees, till and fertilize the soil, and plant millions of seed, to provide trees of the future for California forests." That was the idea behind the beginnings of the nursery, some nine miles east of McCloud. It must have been a good tree growing area, as the dense fast growing stand of second growth and pole sized timber on the site indicated.

Ted and I ran the clearing lines around the forty acres, which had been donated to the Forest Service by the McCloud River Lumber Company. Clearing was started during the fall of '47 by Forest Service equipment and personnel. We cleared, piled and burned thousands of trees, during the cold, wet fall and early winter. A heavy, thick, claylike fuel called Goop was used to burn the still green trees. It was a leftover product obtained from the military after the end of World War II, and burned with white hot intensity, giving off an odor like 4th of July sparklers.

Two prefabricated houses arrived in railroad cars at the Ash Creek side about the time clearing started. They were from Brookhaven, Mississippi, made from southern pine. The growth rings were about ½ inch apart, and many of the panels were warped beyond belief. We stacked and weighted them down to straighten out the twists, but didn't succeed very well, as cracks showing daylight between the panels were obvious during assembly of the buildings. Some "wide putty" and insulation closed most of the gaps, but present occupants say there is still lots of heat going through the walls. They're still pretty good looking houses, though, after nearly 30 years.

During this time, we also were building a dam on Tate Creek to create a reservoir for the nursery water supply. From the dam, some two miles upstream, a ten foot transite water pipe was put together and covered in a ditch to the McCloud River where it crossed on a stringer bridge before reaching the nursery site. Distribution lines of smaller diameters were laid in grids throughout the clearing. From these distribution lines, two inch risers were placed to reach the overhead sprinkler system some six feet above ground. The risers were topped by a hydraulically operated pump which, through cables and rods, moved a rocker arm, and in turn, rolled the overhead sprinkler pipes in a turn of about 45 degrees each side of vertical.

Every five feet on the sprinkler pipes was a sprinkler tip, or about 80 tips on the 400 foot long pipes. The sprinkler pipes were reduced from 1 ½ inches to 1 inch to ½ inch diameter along the 400 feet of line, and the torque from the power of the hydraulic pump created a very graceful, progressively moving spray of water from end to end on the lines. Of course, these tips would plug up with trash in the pipelines, and for a couple of years, before the water system cleaned itself completely, a man would have to walk down the line with a small stiff wire to unplug the tips. This wasn't too bad on warm summer days, but on a frosty spring or fall morning, a guy could get pretty cold and wet.

When the nursery reached a barely operable condition, Karl arrived. He was a hard working, uncompromising, and demanding boss, who knew what he wanted and also knew how much work a man could do in a day. He expected about twice that amount, though. I worked for him about four or five months for each of about four years, and believe it or not, I liked it. Underneath, he was just like he was on the surface. No deceit in the old Swede.

The first day I met him, we went to Susanville to pick up some trees from the small nursery the Forest Service had operated there, and which the new Mt. Shasta nursery was replacing. I had picked up the little International track -laying tractor at the Mt. Shasta shops, loaded it on the truck, and met Karl at the nursery. When we got to Susanville, we had to lift the trees from seedbed, and pack them in bales to bring back to the new nursery to transplant.

"Don't you know how to operate that tractor?" he hollered, as I awkwardly backed it into position to attach to the lifter. "What do you expect when the left steering clutch won't work," I shouted back. He was pretty upset. Getting that steering clutch fixed was the purpose of the tractor being in the shop, and they hadn't fixed it.

We got the trees lifted and baled, despite the mechanical problem, and headed back to Mt. Shasta nursery, where we arrived about 10 p.m. I was soon to learn that time meant nothing to Karl, so the long day was just one of many while working for him.

I learned a lot at the nursery. Pipe laying, dam building, equipment operation, plumbing, carpentry, concrete work, seed extracting and all about raising baby trees. Most of all, though, I learned how to work with people. After working for Karl, I'm sure a man could work for anybody. During the first few months I'd work for different people almost daily. We had the dam building crew, the pipeline crew, the clearing crew and the building construction crew. Each had a different foreman. Working conditions varied from bad to deplorable, with cold, snow, rain and mud for days on end. Someone was always sick, on leave, or had just quit, so each day I found myself working for someone else. This was probably what kept me going, as the job was at least different, if not better, each day. Les ran the carpenter crew, which I liked, because we were out of the mud. Fred ran the dam job, but he was easy to work for, because he would talk for a couple of hours if you got him sidetracked. He really knew his business and I learned much from him.

Snuffy was in charge of the pipeline gang, and this had to be the worst job going. Two of us, usually Bob and I, worked in the bottom of a three foot ditch, coupling the sections of pipe together with special jacks which pulled the collars over two rubber rings to seal the joints. A day in the mud and snow on your knees, astraddle a cold concrete/asbestos pipe left a man in a terrible frame of mind. (Even with tin pants and coat, your legs and rear end were numb and aching by the end of the day.) Why am I doing this? I know I'm smarter than this. Why did I leave that good job? were the kinds of thoughts which entered my mind, but when spring approached, things got back to a rosy perspective, so I hung in there.

When all the construction was completed, I worked as Karl's assistant and Jack-of-all-trades, until Marvin arrived and took over as assistant nurseryman, a job I dearly wanted. I continued doing such jobs as seed extracting, seed bed preparation, seeding and fertilizing through the spring season, and going back to my fire control job in early summer. This I continued for about four years, before getting a full-time fire control job.

The nursery really developed rapidly into a production operation. New equipment, new buildings, and a new cold storage building took shape, and trees were growing like mad, As production increased my jobs became more specialized, mainly in seed bed preparation, seeding, transplanting and lifting stock.

We still were picking sticks, roots, and occasionally large chunks of wood from the ground for about three years after the clearing. Plowing and disking, and even tree lifting would bring woody materials to the surface. These pieces really made transplanting difficult, as the machines were self-guided, and deflection of the guide bar by a stick of wood really put a kink in the seed beds.

Because women were more dextrous, and seemed to be better able to kneel and hunker down in the seed beds, we employed about 20 to 25 local ladies from McCloud and Pondosa to do the transplanting, grading of seedlings, and packing of the bales or crates. During the nursery's first few years we kept the seedlings in the seed bed one year, and transplanted them at the start of the second growing season. There were eight transplanting machines, quite ingenious devices, probably designed for other crops, but adaptable to transplanting young trees. They were powered by a one cylinder Fairbanks-Morse engine of about one horsepower, and traveled the 400 foot long beds in 18 to 20 minutes. Two ladies rode in the backward facing seats behind the drawbar, wedging the transplant seedlings into rubber slots on the vertical revolving wheel. As the wheel turned, the seedlings were lowered into a furrow made by the machine as it traveled, and then tamped into the furrow and pulled from the wheel by two tamping wheels immediately behind the planting wheel. My job during transplanting time was to get on the job thirty minutes early to have the machines watered, fueled, lubed and the engines running plus having the trees ready for transplanting before the ladies got to work.

The transplant beds were prepared by carefully measuring the spacing to the center of each bed from the sprinkler pipes on each side. A taut wire was used as a guide to mark the center of the bed, where a wheeled, one-man push plow made a furrrow for the tongue to guide the machine down the bed. It was very important that the furrow had no jags in it, as each of the seven times the machine traveled the furrow, the jag was worsened, and by the seventh or last trip, the transplant bed was a disorderly mess, with rows of transplants crossing back and forth over each other. Karl could sure get upset over trifles like this. My days were spent running. With eight machines going at the same time, and two women on each machine, there was always something to do, usually at opposite ends of the 400 foot beds at once. Because the ladies rode facing backward on the machines, they never knew if the machine was "taking off toward Jones" until too late, so that's what happened. At the same time another machine needed water, two needed to be turned around at the other end of the row. One of the new ladies was putting the transplants in upside down so the roots were waving in the breeze, while Karl was asking me, "Why can't you keep up?" The morning and afternoon "breaks" at 10 a.m. and 2:30 p.m. were for the girls to stretch, rest, take a swallow of coffee or wine and whatever else they had to do. I wouldn't know, because that was when I fueled, watered, and greased the squeaky wheel. After work, I still had to drain the water from the machines, because nights were cold enough to freeze the blocks in the spring and fall.

Don't get me wrong, those ladies worked hard and fast. Nimble fingers were needed to feed the seedlings to the machine one at a time, at proper depth and at the rate of at least one per second. Most of them helped me turn the machines at the end of the row. Their cheerful disposition and continuous joking made the hours pass faster. There was nothing very glamorous about their jobs, as they wore baggy pants, and shirt or jacket, usually with a bandanna around their heads. The dirt and dust, along with their sunburned noses made them look like peasants. Yes, some of the older Italian ladies had their wine at coffee break and lunch time. "A little Vino don't hurt nobody."

In the fall, we "lifted" the stock from the transplant beds to be baled or crated for storage over the winter. The women also did most of this work, throwing away the trees that were too small, or too crooked, and tying the rest in little bunches of 25 or 50. These bundles were then packed in wet sawdust or peat moss in crates of 1000, 1500, or 2000 depending on the size of trees.

The bales or crates each weighing 50 to 80 pounds were wet down and placed in a cold storage building, where temperatures of 34 degrees to 36 degrees and humidity of close to 90 percent were maintained until the trees were shipped out for planting in the spring.

Storage in the cold room did not end the job until spring, though, as weekly checks had to be made on the trees, checking for mold or mildew in and on the bales, treating with chemicals when found, and eternally restacking the hundreds of crates. This sounds like a pretty bad job, and it was, but the hard work and heavy lifting kept me warm, strong and healthy, despite the dark, gloomy, damp atmosphere in the cold room.

In March and April, suddenly there was a demand for thousands and thousands of trees by the forests of California. Everyone wanted theirs NOW! We would ship truckloads from the cold storage, and begin lifting and crating for shipment those in the transplant beds. This was not a simple operation, because when most forests wanted their trees, there was still four or five feet of snow over the beds at the nursery. Someone should have been aware of the fact in plannng the site of the nursery, which was in a real snow belt of northern California. I'm sure this was the principal factor in abolishment of the Mt. Shasta Nursery some twenty years after its beginning. What did we do about this dilemma? Well, we shoveled, and shoveled, and shoveled. Of course, this covered the adjacent beds even deeper with snow, which later had to be reshoveled into the original location after the trees were removed. We had another method, too, which to me was even worse than the shoveling. Carbon black (like soot) was scattered over the snow to make it melt faster, and it really did, sometimes as much as a foot each day. We poor guys spreading the stuff looked like something out of an old time minstrel show. Black everywhere, except for teeth, tongue, and the white of our eyes. Three day's time and five or six showers finally got rid of the black embedded in pores, wrinkles and fingernails. Just another of the joys of working at the nursery, or for the Forest Service, for that matter.

Not everything at the nursery was bad. It was all hard work, but very pleasant much of the time. A warm spring or fall day, with the beautiful forest surrounding the area, made living and working a pleasure. I was young, tough, and recently married to the greatest girl in the world. Everything was going my way. The variety of jobs, both at the nursery and on the ranger district, made work fun. Always something new to look forward to.

Probably the best thing at the nursery was meeting some terrific people. Karl was a lovable tyrant at times, but someone I'll never forget. He taught me much, or maybe, he helped me to develop certain traits such as patience, understanding and the ability to work hard.

The three assistant nurserymen he had while I worked there probably didn't get along with him as well as I did, but each did his job well. Marvin resigned from the Forest Service after a couple of years, but Doug and Jake went on to better things with the service and both families have remained our friends over the years. Doug, who came to the nursery as a young forester, was very discouraged about his work and future in the Forest Service. He seriously considered going to work in private industry, and had generous offers from more than one company who no doubt could see his great potential. I once told Doug he obviously had a great future with the service if he would just hang in there a few years. It was no great surprise to me when some twenty years later he became regional forester. Doug, Marian his wife, and family have always been fun to be with, never changing as his job status continued to rise. A barefoot young lady, being pulled through the seedbed by a large bird dog while Karl muttered under his breath, is one of the first remembrances I have of Marian. This happened while Doug and I were lifting trees some 150 yards away. "That's just my wife," he said, grinning at the sight.

Only one time did I get into a real hassle with Karl. I woke up one morning at my home in McCloud. Snow was falling fast and furiously, as it can do in McCloud on occasion. About 10 inches were on the ground and the road to the nursery had not been plowed, so I telephoned Karl to tell him I probably wouldn't be there until later, whenever the road was opened. He had lots of work for me to do and told me to come on anyway, which I did. The snow was blinding and the powder flew up over the hood and plastered the windshield faster than the wipers could take it away. Travel was slow, and I'd used about a third of a tank of gas by the time I reached the nursery some two hours late, where the new snow was about 14 inches deep. Karl greeted me grumpily and said something about my tardiness. It was payroll day, and right after lunch, Karl brought my payroll sheet for me to sign. It showed two hours annual leave for being late. For one of the few times in my life, I blew my stack. I'd been working at least a half to three-quarters of an hour overtime every day for weeks without pay, so I wasn't about to take leave for the two hours I'd struggled to get to work that snowy morning. Karl contemplated for a few moments, and changed the paysheet to agree with my way of thinking. Then to my surprise, at three p.m., he sent me home early with pay. The snow continued to fall.

Snuffy and Earl, a real pair of oldsters, were a couple of my favorites at the nursery. They plodded, moved slowly, and piled up more work in eight hours than you would think possible. "You know, Earl," Snuffy would say, "that Karl expects us to do all this work before quitting time." "Yeah, I know, let's do it." And they did, even surprising Karl. They were always assigned to the tedious jobs, where no high ball operation was expected, and accepted these jobs with good humor and enthusiasm. Their cheerful attitudes rubbed off on everyone.

I told Snuffy, who was a bachelor, that I was going to get married. "You're what!" he exclaimed. "Don't you know that the first thing you know you'll have to get a new car, buy a refrigerator, buy a washing machine, not to mention toasters, electric irons and a cook stove. There just ain't no end to what a woman will want." Well, I ignored his advice and got married anyway. He was right, but it's been more than worth it.

Now Snuffy, who was about 65, didn't tell me, but I heard later that he was getting married. I could hardly wait to remind him of his advice to me about all those expenses. When I did, his response dampened all my anticipated glee. "Well, Bob, the gal I'm marrying already has all that stuff."

Abe was a forestry graduate from the East, who had been hired as a district dispatcher. A nicer person could never be found, nor a more gullible one. This, combined with his being the most unsuited man for a Forest Service job made him the laughing stock and butt of all jokes. I don't know how he ever graduated from forestry school, or why he ever went to one. Before fire season he was assigned to the nursery to help in lifting stock, transplanting and all the other physical work. He was soft, untrained, and had no aptitude for the work, but he tried; so hard that it was embarassing to work around him as he struggled to do the job. Everything he did was wrong, and the men rode him unmercifully. Even the women were pretty hard on him, as they could do more physically than he. But, you know, within a week, everyone was solidly behind him because of his cheerful personality and sincere appreciation to those trying to help him. When time came for him to go to the district job, which was only about two weeks after he began working at the nursery with us, we gave him a "going-away party" during a coffee break and even Karl participated in it. He was delighted and surprised at our actions. I wish there were a happier ending to the story, but the last hour of the day he got his fingers caught in the heavy doors of the cold room, mashing them pretty badly, though nothing was broken. He then worked only part of the summer at Dunsmuir before resigning and going back East. Abe, if you ever read this, I know that I and all your acquaintences at the nursery sincerely wished you well in whatever you did back East.

TREE PLANTING

My own estimate tells me that I've personally planted 100,000 trees for the Forest Service, and have had a hand in planting another 150,000. Right or wrong, that's lots of planting. Two steps, and then bang the hoedad into the ground at arms length, pull the dirt out with the blade, stick a tree into the hole, push the dirt around the roots with your hand, then stand up and pack the loose dirt in the hole with your heel as you start the procedure again. Keep this up at the rate of once a minute and you'll plant 480 trees in an eight hour day. Sometimes slower, sometimes faster, depending on the soil, rocks, brush and terrain. Anyway you look at it, it's a tedious, backbreaking job, and leaves you ready for a good night's sleep. Some grow, some don't, depending on soil moisture, insects, rabbits, deer, pocket gophers, dry winds and condition of planting stock. Tree planting is big with the Forest Service, and rightfully so, as many thousands of acres of non-productive brushfields cover the mountains of California and the West. Not all is capable of producing timber, and the prospective timber producing areas are carefully studied to determine feasability of planting. Soil depths, chemical makeup, rainfall and terrain, among other qualities, are determined before planting is attempted.

An interesting tree planting area was the eroded, barren slopes of the rugged hills around Shasta Dam. It was during the middle and late forties that I was a member of a crew working and planting this area which had become denuded from the toxic fumes of the Kennett Copper Smelters of years gone by. Each morning our crew of about ten men would leave the Turntable dock by boat to go the several miles to the planting area. Leaving camp about 7 a.m. it would be about 8:30 or 9 a.m. by the time the trees were stowed on board the outboard motor boats and we'd completed the trip to the devastated area. The trips made the day worthwhile. At one place we passed by an otter run or slide, and most mornings the otters would treat us to a frolicking show as they climbed the bank and slid down the muddy run into the lake. On occasion we'd see deer swimming across the lake and we would carefully go around them at a distance great enough not to frighten them in the water. They are equally graceful animals swimming or on dry land.

When we got to shore the work began as we unloaded the trees and our water and lunches from the boat. The steep climb up the bare slopes took lots of energy even before the planting began. At the elevation of about eleven hundred feet, the days became hot soon after sunrise on these east facing slopes, even in March and April. It appeared to be a hopeless task to ever get trees to grow in the rocky, eroded soil, but we planted and they grew. Now, some thirty years later, the slopes have been stabilized, and a beautiful young forest of pines is growing. A successful plantation of several hundred acres now covers the once desolate area.

Other plantations were not so successful. I was in charge of planting ninety thousand seedlings in a grassy flat on the McCloud District in 1949 or 1950. The terrain was flat, the weather perfect, trees available right out of the nursery less than 15 miles away, and we were using machines. Since the days were long, we ran two shifts, from daylight 'til noon, and noon 'til dark. The machines were made by Lowthar, and towed two-at-a-time by a tractor. The planter rode on the machine on a seat between two tilted wheels immediately behind a plow point and coulter wheel. The coulter wheel cut through the sod and litter, the plow point opened the furrow, where the tree was placed, and the tilted wheels packed the dirt back around the tree roots, resulting in a very well planted tree.

Things went without a hitch. The trees were planted in record time, and three weeks later had started growing when we first saw the grasshoppers. They were everywhere. You could see them chomping away at the needles on every tree, sometimes two to a tree. Within three days there wasn't a needle left on a stem. We found fourteen live trees later in the summer.

We tried a new system for tree planting in an old burn northwest of Helena in Trinity County. It was a team effort, with one man using a gasoline powered post-hole digger to dig the holes. The "Little Beaver," as the digger was nicknamed, could keep ahead of several planters, and really made a neat, clean hole, with a mound of well worked dirt deposited around each hole. It speeded up planting to an average of 700 to 800 trees per man in an eight hour day, and the psychological effect of the motorized device bolstered the morale of planting crews. Running the "Beaver" was the hardest part of the job, but everyone wanted to operate it, so we would swap operators about every hour, at refueling time.

Some of the most successful plantations on the forest were planted this way, both on the Weaverville and Sacramento ranger districts.

DISPATCHING—WAS NOT MY CUP OF TEA

As a bachelor, I could get along on the meager pay of a GS-4 in 1948. But a year or so later, with a new bride, and not long after, a baby on the way, I had to find a better job. Well, the job I got wasn't better, it just paid more. I went to work as a scaler for "Mother McCloud," a name given the McCloud River Lumber Company by its employees. For three or four months I scaled logs at Pondosa, during a winter I'll never forget. Snow was 10 feet deep and we were still logging. Railroad logging was coming to a close, but that's what we were doing. The old oil burning steam engines would plow the tracks, bringing in loads of empty flatcars, while a fleet of dozers was keeping skid roads open, and plowing out and skidding logs covered by the almost continually falling snow. The days were long, from five a.m. to five p.m. or later, and I longed for the old job with the Forest Service. "If only they paid a decent wage," I thought, "I'd go back to work for them."

It took twelve feet of snow to finally stop that logging show. The old engine took a run at a drift of snow on the track one morning, and found itself high centered on snow. The drivers were churning air and snow about an inch above the rails. Ed, the woodsboss who had worked and fought snow for weeks on end finally gave up. "That's all, boys, see ya' in the springtime," he said and we headed for town. Unfortunately, for me, the road to McCloud had snowed in, and was to remain closed for several days. I stayed in Pondosa two days until the road to Burney was opened, and I went home the round-about way of Burney to Redding to McCloud. When I got there, the Forest Service asked if I'd like another job with them, with a promotion, in the dispatcher's office at Mt. Shasta. I jumped at the offer, though I greatly appreciated the work given me by "Mother McCloud" in the interim, and I think the association with the private sector of forestry was to my advantage in later years with the Forest Service. I know that my acquaintanceship with Bob M., Bob L., Ted, Ed, Mr. G., and others with whom I worked were lasting and worthwhile.

The dispatching job was interesting in that it was new, and I was working with a different group of people. Merv was my immediate boss—or was it Ralph? I never really was sure, but both were fine to work with or for, whichever. Merv always had an immediate answer for anything and made decisions without hesitation. Good decisions, too. He could store more information in his head and have it at the tip of his tongue in seconds, than most people know in a week. And, he had a note of authority in his voice that got things done. When Merv went after something, whether a commercial airliner, or a wheelbarrow, he got it. In my job as his assistant, I learned and did a lot, but never developed his mastery of dispatching.

Now, Ralph was a different personality entirely. He talked and thought at a slower pace, but with a finality that convinced people he knew what he wanted and was going to get it. As fire control officer of the Shasta Forest, he had a tremendous job in making an efficient fire organization, and getting cooperation from the Southern Pacific Railroad, which was giving us nothing but trouble from fires during these years. In our spare time, Ralph and I ran a Boy Scout troop in Mt. Shasta.

When California was burning, dispatching was fun.

Ordering airplanes, buses and people, along with tons of supplies and equipment, kept the adrenalin and interest at a high peak for me, though I sorely missed getting in on the live action and excitement of fighting fires. It was the periods between fires that I could hardly stand. Coloring maps, making lists, checking job applicatons and posting atlases "drove me up the wall," and that expression wasn't even in use in 1953. Well, I learned a lot, and held up pretty well for two years, by sneaking in an off-forest fire or two and doing some nursery work and engineering field work pre and post-season.

When Mr. Jones and Ralph asked me if I would be interested in a district F.C.O. job at Fall River Mills, I said yes, even before they could get the location out of their mouths. Except for the good relationship I had with Ralph and Merv, plus Flossie and Graycie helping during critical fire situations, I probably would have gotten out of fire control after two years dispatching. It just wasn't for me, though I thought I'd like to go back into it after I got older, like maybe 40 or 45. On occasion I did help Merv, Bud, Rip, Ted, Larry and Ralph R. in later fire seasons; enjoying it thoroughly with the knowledge that it wouldn't last forever.

SMOKEJUMPERS

A lot of terms describe smokejumpers—brave or foolhardy; elite or snobbish; daring or showboaters; glamorous or pampered. It depends on who you're talking to. No doubt, they're all of this and much more. It takes a special man with lots on the ball to become a member of this rather exclusive group in the Forest Service, and a little bit of envy or jealousy accounts for some of the less flattering adjectives used to describe them by some. In reality, they're a proud, tough, well-trained group of firemen, who just happen to travel to fires by a different method than the rest of us. They're good, and they know it. From my viewpoint, "that ain't bad." Many top-notch firemen in our region are graduates of the jumper program and have gone on to top fire management jobs all over the country.

With the present and improving transportation system in Region V, I'm not sure the Forest Service can afford the luxury of the smokejumpers for many more years in California. There are few places now-a-days which can't be reached by ground travel within an hour of the time it takes the jumpers to reach it. This is especially true on the Shasta-Trinity, though this is not necessarily so in Region VI or the Montana-Idaho country. Helicopter use has also reduced the need for jumpers. Don't get me wrong, I've made dozens of jumper requests over the years, and so long as they're available, would continue to do so, with the feeling that dispatchers should have the final decision where they should be used when a priority situation arises, rather than use them on a first call, first serve basis.

Watching jumpers leave the jumpship is a beautiful sight, especially from a recon plane circling above the jumpship. The sun shining on the bright orange and white parachute canopies against the green forests below makes a colorful scene. On one occasion, the Thurston Peak fire of the Big Bar District, I watched fifty jumpers as they made initial attack on the fire. It was the most jumpers on one fire in the history of the region, and was a spectacular sight to watch the cargo and personnel chutes drift and settle into the timber and brush around the seventy acre fire in the beautiful and rugged Trinity Alps wilderness area. I was air boss on the fire, which was an impressive air show, with air tankers, smokejumpers and helicopters working simultaneously on the otherwise inaccessible fire. After guiding and directing the air show for about three hours, I took over as line boss on the ground. It was a tough show for everyone, and after a trip on foot around the fire, I could really appreciate the tremendous job done by the jumpers who worked through the night without rest until the final lines were tied in. After the control lines were completed, and with a couple of hours sleep, it was time to begin retrieving the two tons of gear that had been dropped in with the jumpers. This is where the job changes from glamour to drudgery. A definite lack of enthusiasm was apparent as cargo was retrieved from the steep slopes and canyons around the fire, but they did it in a very businesslike manner before taking another well deserved rest.

I really got acquainted with jumpers on the job at Weaverville. We had two jumpers go in on a small, easy lightning fire in the upper north fork of the Trinity. The fire was just above a trail, and about ten miles from the road end at Hobo Gulch. I'd figured the jumpers would have the fire out by the time my son, Doug, and I got there with a horse and mule. Our timing was perfect, as we met the two jumpers just as they reached the trail with their packed gear. They didn't know we were coming, and had planned to pack their 90 pound loads out themselves, so were very pleased to see a pack animal. Just for fun, I loaded one of their packs on my own back, and let me tell you, the way I staggered around with that load, I don't think I could have packed it out the ten miles.

There were three jumpers on a small fire I had up Canyon Creek in 1957, and after landing, one of the three was hopelessly hung up in a large tree. His two "friends" decided to leave him there while they put the small fire out, which they did while he hung helplessly in the tree. He was a smoking smokejumper by the time they got back to him a couple of hours later to help get him down.

My first experience with smokejumpers was in 1952 while working as assistant zone dispatcher with Merv and Ralph. Dispatchers did a little of everything in those days, including flying recons for many of the districts. On one recon over the old Redding District, I'd discovered a fire near the top of Claiborne Peak which is real jumper country—high, inaccessible and dense. The fire was barely showing a little smoke at the base of a sugar pine, and was in a place easy to describe to the jump plane, but the jumpers just couldn't find it. Now this was a Ford tri-motor jump plane from Missoula, with a load of nine jumpers, all ready to jump if new fires were found while they were in the air. The plane came back to Mott to report that they could not see a fire where I'd described it to be, so they invited me to come with them to show them where it was. Well, flying in the machine was a thrill. A six foot square gaping hole was where the door should have been, and the corrugated metal skin rattled and groaned as we flew over the area where I'd seen the smoke earlier in the day. But no, there was no smoke showing now. I could recognize the spot but couldn't see any sign of a fire. The jumpers began to needle me and in jest (I think) threatened to throw me out the door if I didn't find that fire, to which I replied, "It will take more than the nine of you to get me through that hole." We continued to circle, and finally one of the jumpers shouted, "I see it, there is a wisp of smoke there." None of the rest of us could see anything, but since I was positive there had been a fire there earlier, two of them jumped. The story ended well, they found a smoldering fire in the duff under the sugar pine and had it out in record time.

MORE FIRES WORTH TELLING ABOUT

The Sacramento Ranger District has a past history of many man-caused fires per year. In 1960 I arrived as the new fire control officer in Mt. Shasta. A man in that job really should be an overseer, organizer, planner and expediter, which I was when the occasion demanded. But, if the occasion did not demand those needs (in my opinion), I would grab a shovel, hose, ax, or some tool and go to work with the crews. There were always radios on the fireline, and on small fires it was no trouble for someone to find me if needed. John, my boss, was not quite of the same opinion but had given me a free rein in fighting fires, so I could do what I thought best. He never really felt comfortable with my statement to him, "If you don't hear from me on the fire, everything's O.K., and if we are having troubles, I'll be on the radio."

We had a small railroad fire one day, which obviously was not going to be a problem, so I left my pickup with a shovel to work with the crew. Not because they really needed me, but there's no better way to have good working relations with men than to physically do the job beside them. Suddenly I looked up to see a husky young crewman beside me with a radio in his hand. "Does someone want me on the radio?" I asked. When he answered no, I asked why he wasn't working with the rest of the crew and he very nervously answered, "Mr. W.—is up on the road, and he says, 'You get down there with this radio and don't let Bob Gray out of your sight.'" Well, I took the hint, and carried a radio after that, that is, until John was transferred.

Some fires were fun, especially those on the bank of an alpine lake. Bucket brigades with hard hats can make quick work of a fire. Keeping the crewmen out of the lake can be a challenge, so we didn't try. Instead we just took turns swimming and fire fighting. The unlimited water supply even made the mop up fun. An occasional diversion like this makes the normal nasty job of firefighting more bearable.

Bernie and I discovered a small fire burning in deep duff at the edge of Canyon Creek Lake. It was smoldering in a circle about three feet across, so we just shoveled it into the lake before setting up camp for the night. Instead of going out like it should have, some of the smoldering duff floated across the lake and yes, it caught some driftwood on fire on the other side of the lake from the camp. This time we put it out as professional firemen should do. There's nothing like taking shortcuts to get a man in trouble on a fire.

The Pilliken fire on the El Dorado was an interesting fire, but there's just one thing I keep remembering. I got run out of the shower by three young ladies. I'd come off the line early for a strategy briefing, so decided to take a shower. It was a Jerry-rigged set-up—about six shower heads in a canvas closure and I had it all to myself. Then I heard a female voice, "It says Line Boss on the hat." She was obviously looking at my pile of clothes and hard hat by the entrance to the shower.

"Well, let's go in, there's plenty of room," another high pitched voice responded. I frantically rinsed the soap off, and was streaking for my towel. I never made it; it was with my clothes. We met half way. I dressed outside, listening to giggles and splashing water from within the closure. I delivered an ultimatum to the camp boss. It read, "There will be scheduled showering time for males and females." I hate being giggled at.

Some of the nicest and most memorable experiences on fires were not at the fires but enroute to and from. I never was much to remember whether I went somewhere in a Ford or a Continental, a Chevy or a Cadillac. It's the same with airplanes. Maybe a Piper Cub or a Concorde, all I can remember is little, middle size, or big. Anyway, it's not the kind that counts. My first trip to the Southern California mountains was at night. I'd never taken many airplane trips anywhere, and was disappointed that it was to be a night trip. I found a window seat though, and tried to navigate myself by counting towns and cities as we crowhopped from Montague, Redding, Red Bluff, Chico and Sacramento where we finished our load of overhead. Then on to Los Angeles. From Sacramento on I was lost. I still have problems remembering, does Fresno come after Modesto, or is that Stockton down there? The cities are beautiful at night. Colorful, symmetrical, sprawling, strung out, scattered; all connected by strings of lights as cars travel the freeways and highways. Bakersfield is the last big cluster of civilization at the south end of the San Joaquin Valley. Then a bright line of headlights over the Tehachapis via the Grapevine with nothing but blackness on either side until Castaic Junction shows up. From that point on, the lights of the hundreds of communities of the Los Angeles Basin fan out, blend together, glittering like diamonds as far as you can see in all directions. This was all new to me in the mid-forties. All the old-timers were sleeping or talking, paying no attention to the lights below. It was September and probably Friday night; I could see lighted baseball diamonds and football fields and other airplanes zooming by. Suddenly, someone says, "There's our fire, off to the left." Sure enough, there was a solid curved line of fire on a mountain with billows of smoke showing in the moonlight. "Pretty impressive fire," someone said. To me it didn't look so bad, but then, I was new at the game and didn't realize those flames were fifty or more feet high, and was looking at them from ten miles away.

This was my first trip south, but you know, I never failed to enjoy any of the following trips—day or night.

The Forest Service made a bad mistake on the Cleveland Forest in about 1946. They turned loose 30 or 40 Northern California overhead on uptown San Diego after releasing them from a fire. Now, I'm not much for drinking, but not all my friends are teetotalers, and among these forty "friends" were thirty-nine non-teetotalers. For some reason the group stuck together after dinner, as they wandered the uptown section of the city, not far from our motel. After hitting a few bars, some of the boys had one or two too many, and before long, most of them had overindulged. Since I had remained sober and a little bit disgusted with my friends, and was afraid the whole bunch were going to wind up in jail, I decided to go to a movie. "Oh no, " Lyle said, "we need you to keep us out of trouble." They promptly elected me treasurer of the rowdy bunch, and pooled their money for me to pay their bills as they left each bar. I must have had $400, which was lots of money in those days. There were some worried looking tavern keepers as they ordered en masse and said, "That guy over there will pick up the tab."

We survived the night with no arrests or fights. When it was over, I still had about twenty dollars in the pool. "Keep it," they said. "You earned it." I agreed.

Transportation varied in class and accomodations from the 40's to the 70's. Sometimes it was a stripped down Forest Service plane used for smokejumping and cargo dropping. For instance, the "Vomit Comet," a shell of an old DC-3, nicknamed by the smokejumpers who used it for years. It had wooden benches along each wall, and the heaters never worked right. Other times it may be a plush commercial job with stewardesses, champagne, and a free lunch.

Forest Service people were very critical of some of the charter services, describing them as the "Mercy Airlines," the "Pathetic South West," or the "Air Sick West." Anyway, we always got there and back, despite a bouncing landing or two enroute, or a near miss with some headon plane flying from the opposite direction and at the wrong elevation.

One fun trip was in a "de Havilland Otter." About fifteen of us left Southern California in it after an Angeles fire. The overhead team was from all up and down the state, so we landed everywhere. Fresno, Sonora, Placerville, Reno, Quincy, Susanville and finally Mott, where I was the last passenger. Each time the load was lightened, we would land and take off in a shorter and shorter distance. I'm sure at Mott we landed like a pancake and didn't move over 10 feet after hitting the runway, and the takeoff wasn't more than 20 feet after I got out. The pilot had fun and so did we, as he demonstrated the performance abilities of this very maneuverable ship. He didn't even scare the jackrabbits off the Susanville strip.

TALES ABOUT PEOPLE

Les worked for engineering out of the supervisor's office in Mt. Shasta. He was one of the old guard type; tough, dedicated, and always cracking jokes in a cynical kind of way. He knew those "damn cigareets" were going to do him in, but "a man's gotta have a few vices," he said, as he wheezed, coughed and puffed away.

One Friday afternoon he finished some bulldozer work near Esperanza, where he left the Cat for the weekend. Monday morning he started walking in to Esperanza from the highway to get his machine. About 18 inches of snow had fallen since Friday, but since it was only a mile to hike from the plowed highway, he took off. On reaching the spot where he left the dozer, all he found were tracks heading up the hill. After a little contemplating on what to do, he set out on the trail of the Cat, because the tracks were fairly easy to follow. He walked for a couple of miles and began to have second thoughts about getting the Cat. No telling how much further it was, so he headed back to Esperanza. It was getting cold and dark.

In the meanwhile, it had gotten long past quitting time and Les' wife had called to say he had not returned from work. Paul and I set out to find him. As we neared Esperanza we were very apprehensive as to whether we would find him alive. Now, Esperanza was an old siding on the McCloud River Railroad, and there were several abandoned houses where the section crew used to be quartered.

Standing in the door, with smoke billowing from behind him, was Les, with a big grin on his face. "I figured someone would be here 'fore long," he said, and added, "I just plain ran out of steam." The smoke was from a fire he built in the middle of the floor to keep warm while he waited. The building to the right was already a smoldering pile of embers. He had burned it a couple of hours earlier. There were still several more buildings to go, so he wasn't too worried. Les lived quite a few more years, in spite of his wheezing and coughing.

ED & A BUCK

Ed was a fire prevention technician at Fall River when I went there. Now Ed was famous for telling tall tales, but after working with him for a couple of years, I decided he was just one of those lucky people who always had interesting things happening to them.

The flats around Burney and Four Corners are a winter deer range, because the snow never gets too deep, and goes off early in the spring. Ed and I were marking some dead and dying pines in the area for a salvage timber sale near Four Corners. We were working strips about 50 yards apart, and keeping track of each other by sight or calling back and forth. Ed called, "Hey Bob, come here, I need some help." I thought maybe he couldn't decide on whether to mark some trees or not, as I walked toward him. When I got there, he was peering around a tree at something. The something was a huge buck with blue antlers. I stopped and watched as the buck followed Ed around trees and thickets, while he squirted blue paint at him. That buck was definitely trying to pick a fight, and Ed was equally determined to keep away from him. The whole front end of the deer was dripping blue paint and he continued to follow Ed until he saw me, when he decided he couldn't cope with two humans at once. I don't know yet why the buck acted that way, and only one other time have I seen a buck show hostility to people, and he was obviously sick and foaming at the mouth. Nothing at all appeared to be wrong with Ed's buck.

A few days later while working in the same area, we sat down on a log for lunch. Never hungrier, I opened my brown bag, and stared in disbelief. It was the garbage bag. I'd picked up the wrong sack on the drainboard. Needless to say, Ed thought it was hilarious until I informed him we would be sharing his lunch.

LOST HUNTERS

Local sheriff's offices are responsible for organizing search parties, with the Forest Service serving in any way they can help, often providing personnel, communication and leadership for search parties.

On one occasion while I was in Weaverville, the sheriff asked if I could fly as the "eyes" for their recon, since I was familiar with the rugged north fork of the Trinity River. Two hunters were two days overdue in this area. This was the first time I had ever really looked for anything other than smoke while on a recon. No, we didn't locate the hunters, but, much to my surprise I saw things I'd never seen before. Never in my many recons had I seen a bear, and only one or two deer, but this day was different. Lonnie was the pilot, and we saw at least 12 bear and several deer, as we were looking for moving objects, specifically men. Of course, our low elevation and slow flying made them more visible than they would have been at normal flying altitudes.

Though we failed to find the hunters, they saw us in our many passes over the North Fork. How do I know? Well, they found their way to a road several hours later and one of my employees picked them up. The sheriff and I met them when they reached town, hungry and tired, but otherwise O.K. My conversation with one of them was like this:

"We really covered that country but didn't see you. Why didn't you build a fire so we could see where you were?" I asked.

"Well," he answered, "we didn't think about that, but we did shoot at you, trying to put a hole in the airplane wing, so you'd know we were below you." Now, I still don't know whether he shot at us or not, but he wasn't smiling when he said it. Neither did Lonnie smile as we searched for a bullet hole in his plane. If he shot, he missed.

FALL RIVER FOLKS

A new job, a new place and new people always leave vivid memories. Moving to Fall River Mills was no exception. My goal with the Forest Service was to become a district fire control assistant, fire control officer or fire management officer, any of which are the same, depending on the year of incumbency. My first was in 1954, after being a lookout, district dispatcher, crew foreman and assistant zone dispatcher. I went to Fall River with high hopes, enthusiasm and a little apprehension as to how I'd be received, and how well I could perform. After being told by the forest supervisor, "Don't get involved in the local schools. Don't get involved in the local church. Don't get involved with the local newspaper," and a dozen other "Don'ts," it was a wonder I got the job. Because my response was, "If I can't be trusted to be myself, and choose who I get involved with, and how my involvement will be, don't give me the job," which I desperately wanted. Well, I got the job, and managed to stay out of trouble, despite my involvement with all the taboo civic functions.

Meeting my co-workers, some of whom I already knew slightly, had its interesting aspects. Sam, the foreman at Wiley Ranch, was a cynical looking older individual. His greeting to me was, "I think you ought to know we all think Ed should have gotten your job." "I appreciate your frankness, Sam," was the only response I could come up with after so sudden a jolt. Our working relations got much better as time went by.

George, the station foreman, a recent immigrant from New York City, greeted us warmly, and though he lacked Forest Service type experience, he ran a good crew, and we put out lots of fires together the next two years. His handiwork around the district always had a built-in characteristic of old George. For instance, he used copper tie wires on telephone pole insulators with galvanized wire, because the contrast looked "classy."

A visit to George and Bert's home was sheer hospitality. Good coffee, good company, a serene setting in the pines, with a coyote howling in the not too far distance was typical. Their home, too, showed the unique George touch. A leather latchstring to open the door, a cow bell for a knocker, and a barn-shaped guest house were all part of the scene.

A couple of weeks after my arrival at Fall River, Donna, on Soldier Mt. Lookout, called late one evening to report that "Mayfield is at it again." "What are you talking about?" I questioned, not yet having met Mayfield Machache, an old, old Indian living at the foot of Soldier Mountain. It seems that once or twice each summer, since "way back when," old Mayfield cleaned up around his shanty by burning all the debris—a commendable undertaking, but he never bothered to obtain the required burning permit from the Forest Service.

"Guess we'll have to write him up," I told Ed as we drove out to check his burning. "Now, Bob, don't be too hard on old Mayfield—after all, he and his ancestors were here before any of us, and he ain't about to change his ways because of you, me or the whole Forest Service."

When we arrived, the old Indian was surveying his job. Well done, he seemed to be thinking. I had to agree myself. Nevertheless, I figured to give him a lecture, if not a ticket. I did neither. As we approached him, he stood squarely in front of us, arms folded across his chest. "Machache's land, Machache's fire—Machache take care of it—don't need Forest Service."

"Let's go home, Ed. No sense in creating an Indian uprising."

Ed grinned all the way back to town.

I'm not really sure this was the best way to handle the problem, but it was probably better than the handling of the "confrontation," "uprising," "rebellion," or "incident" of the early 70's at Four Corners between the Forest Service and the Indians of the Pit country.

Doc was the Burney Falls foreman, another import from the East, totally untrained and inexperienced in Forest Service activities. He was a willing learner and a good worker and worked well with young crewmen.

His lack of Forest Service knowledge showed up one hot, dry summer day. I'd taken him along to work the Big Bend phone line. After giving the crew a lesson in climbing and tying insulators, I went back to where Doc was working. Here was Doc, up the tree, working away and smoking a cigarette. Beneath him was the worst pile of dry needles and litter you ever saw. No one had ever told him that careless smoking causes forest fires.

The whole "G" family worked on the district. Winnie was district clerk, Bill was tank operator at Burney Falls and Bill Jr. was on the fire crew at Fall River Mills.

Bill Jr. had only been working a few days when he asked, "Bob, may I have the 4th of July off?" Well, in those days, no one had the 4th off. It just wasn't done, especially among fire control people. I told him, "I'm sorry, Bill, but it's times like the 4th that we need you most. Do you know that I've never had the 4th of July off in all my years with this outfit?" "What if I just don't come to work?" he asked. "Just don't bother to come on the 5th either," I replied. Bill took his day off and joined the army a few days later.

Two weeks later, on the Los Padres, I was talking to my crew bosses about how to handle their crews, a bunch of army recruits, when a voice from the ranks called out. "Hey, Mr. Gray, I'll bet you didn't know I'd be working for you again in two weeks!" It was Bill. We got a big laugh out of the situation, as I gave him the latest on his family and the Forest Service.

There was Mel at Pondosa, a local of Indian descent, and probably the best fireman of all the crew foremen. He was a woodsman, hunter, and the most unpredictable employee I'd encountered to that time. I learned much of the country while working with Mel, and also improved my woodsmanship just by watching him as he hiked, looked and listened his way through the forests.

Sid, at Big Bend, knew the country, his way around, and best of all, he had rapport with the peoples of the Lower Pit River. It wasn't always easy to deal with some of the Indians, or the ranchers in the area, but things went better if Sid was along. Sometimes his casualness in fire fighting unnerved me, but thinking back, he never let one get away. He really had things well organized in the Bend, with cooperation from all the citizens, a cross section of which included outlaws to highly professional engineers.

While attending a tunnelling-through "picnic" celebrating completion of one of the power projects, some highly inebriated construction workers took over the radio in my Forest Service pickup, gleefully broadcasting obscenities and other inelegant remarks about the Forest Service. I'm sure, without Sid, the revelry and ribald humor would have continued until the sheriff stepped in, but Sid, in his homespun, soft approach persuaded them to cease and desist, without their even knowing they'd been silenced.

Johnny was a challenge, to put it mildly. Hyperactive, full of spirit, and raring to go. In those days, I prided myself that no one working for me was going to produce more work than I could. Johnny not only could, but did. Whether fighting fire, digging ditches or pruning trees, he did more than anyone. His problem was that after work, his energy and activity increased, to the point that his tired old foreman could only take about two weeks at a time of him. Sam once told me, "Bob, you have to do something about Johnny." "What's the matter," I asked. "Is he not working?" "Oh, no, it's not that, he does twice as much as anyone else. I just can't stand him after hours."

Johnny continued working hard years after, and I think he finally got rid of some of the surplus energy. Until he got out of fire control into special services in recent years, there was no one better on the fire when the going got tough.

Ned was the one person I had an "in" with on arrival at Fall River. A year or so earlier, he had been looking for a fire along the West Lava Rim Road, and the directions given just weren't understandable to him. I butted in on the radio because I knew something about the fire location that his dispatcher did not know. With this little bit of assistance to Ned, he thought I was a pretty sharp guy, so we hit it off from the beginning of my arrival at Fall River.

Ned was older than most and had worked on the Pit District for over 30 years when I arrived. His job was fire Cat operator in summer, road construction and maintenance in spring and fall, and like most of us, a jack-of-all-trades at all times. He had been an equipment operator since the first bulldozer was used by the Forest Service. The new-fangled ways of the Service were getting to him and a deteriorating hearing problem was slowing him down, but he was still a man to see when practical answers were needed. Having known Ned was a real plus in my experiences. Last time I saw him, he still had the look of the Marlboro man, even at about 80.

Less than two years after moving to Fall River, I accepted a transfer to Weaverville. These two years were great years for me. My first F.C.O. job had been a challenge, and a pleasure working with fine people and meeting new friends. The country was beautiful, and wildlife abounded. While contemplating the transfer to Weaverville, I drove to the Timber Crater area. In the wide, shallow crater was a large herd of mule deer browsing on the new spring growth. They paid little attention to me as I drove through the area we had recently pruned and thinned as part of a timber stand improvement program.

On the way back to town I looked into the waters of Fall River to watch several large rainbow trout feeding in the clear, cold stream. At the same time a band of Canadian honkers flew low overhead to settle in the stubble field nearby. Only a few days before, Betty and I had watched a herd of about 200 deer on the slope of Big Valley Mountain, along with a small band of antelope. All this almost persuaded me to stay in Fall River Mills, but some stronger persuasion in the form of a promotion and raise changed my mind.

Don't you hate people who sell out for a few lousy bucks?

It was 1956 that the Pit and McCloud districts had a thunderstorm which I will always remember. Air tanker use was still in its early development stage, and I had gone to Fall River Mills from Weaverville to coordinate efforts and forces in suppressing the two dozen or so fires in the Papoose, Stud Hill, Powder Hill and Six Shooter areas. It soon developed that I was running the air show, as observer, locator and strategist in

Air Tanker, 1956

determining which fires would be targets for the air tankers. Ground crews were scurrying all over, picking up fires as they went, but obviously the ground forces were going to be inadequate to man all the fires during the day, and smokejumper requests were bogged down in backorders and priorities. We had ordered and received six or seven converted crop duster planes from Willows to work out of the Fall River Airport. These were souped up Stearmans; a beautiful bi-plane, open cockpit, with a 50 gallon slurry tank built into the passenger cockpit. As air attack boss, I was flying in a Cessna 170, with no communications to the Stearmans. Our technique was to take off from the Fall River airstrip and fly in an oblique formation to the fire area. After selecting the most threatening fire, or the one showing the most potential, I would signal by hand out the window to the first ship behind me. He would make a practice run at the target and then a final run, dropping his 50 gallons on target, nearly always with deadly accuracy. The remaining planes would circle above until I decided whether we needed more retardant on that fire or should continue to the rest. This system continued until each ship had dropped its load. Immediately after dropping, the Stearman would head back to the airport to refuel and refill the retardant tanks. I don't know whether this delaying action was the reason for our success in holding all the fires to small acreages, but the experience is most unique to me, and the competence and effectiveness of the pilots was remarkable.

The leader of a World War I air squadron must have had the same feeling as I did in guiding those vintage ships to their target. Of course, I was in a modern closed cab airplane and wasn't getting shot at, but I could glance back at the bi-planes and see the leather helmeted and goggled pilot's head sticking out of the cockpit. One pilot even had a billowing white scarf.

Since that time, air tankers of 3,000 gallon, or 60 times the capacity, have been put in operation, and the old Stearmans have long since retired or returned to their crop dusting duties.

As air tanker use became more commonplace, I would always remember the "squadron from Mendocino" as my favorite air tanker experience.

TRINITY COUNTRY

The Shasta and Trinity national forests became one in 1954, known as the Shasta-Trinity National Forest, an action I'm still wondering about. Both were large diversified forests, with headquarters in Mt. Shasta and Weaverville. Reasons given at the time were to increase efficiency, to eliminate duplication of jobs, and a host of other vague considerations. I'm sure the real reason was political, like moving a forest headquarters to Redding, to justify a bigger and better north zone service center and to satisfy some high level curiosity about administering a Super Forest. Well, regardless of any reasons, the consolidation took place, with some hard feelings among the townspeople of Mt. Shasta and Weaverville and within the ranks of Forest Service people affected by the big shift. Some twenty-two years later, there are still a couple of "Trinity National Forest" entrance signs still in place on the Trinity side, and some on the Trinity side refer to the "Trinity-Shasta" Forest. On one fire, the timekeepers insisted on a shorter version of the forest name, using the first two letters of Shasta and the last three letters of Trinity.

In April of 1956, I transferred to Weaverville from Fall River Mills. A devastating flood had ripped through the country in December, and had washed out roads, bridges, and large sections of trails. Repairing the trails was part of my initial job, with special financing from flood monies. As I look back, some of the workers we had employed were pretty bad and the results were the same. The transition from spring to summer went rapidly, but with the same personnel becoming my fire crews. Now, they weren't all bad, but enough fit that category to affect the crews. We made it through the summer without losing our shirts, but even a year or so later, the quality of the crews left something to be desired, as Tom and Rod, the crew foremen, did the best they could with what they had. Had it not been for good prevention efforts by Homer and Bernie, we might have had lots of troubles as we went through the fire seasons of 1956 and 1957. Most of the problems with the crews involved personalities and lack of experience, because on the occasions when the chips were down, they did the job and caught the fires small. Tom, the Junction City foreman, later became the F.M.O. at McCloud and is now one of the best in the region.

Weaverville was booming when we moved there. The Trinity Dam construction was beginning, and the town was crowded with construction workers and the Bureau of Reclamation people who were administering the Trinity River project. You'd think all Federal employees would be alike, but not those folks. They shared our office, but didn't bother to speak to us when we met in the halls. Well, the Forest Service has always been a close-knit outfit, very friendly and close in both work and socializing, so I determined to bring them up to our level of friendship. Whenever I'd meet one of them in the hall, I'd confront him, zig when he zigged, zag when he zagged, until they started greeting me with at least a "Hello, Bob." Before long everyone was speaking in the halls and even on the streets uptown. Alice, our clerk, said one day, "Wonder what's happened, they're becoming downright civil to us."

If my figures are right, there are about 17,000 acres covered by the Trinity Lake (officially Claire Engle Lake), and it all had to be burned during the logging and clearing operations. That's enough to cause lots of sleepless nights to a fire control officer, and I spent much of my off duty time checking these burning operations, writing permits and just talking to the clearing contractors. With luck and good prevention work by Homer, we never had an escape from any of this burning. Most of the timber in the reservoir area was pole sized Douglas fir, with minor volumes of larger saw timber. A steady stream of overlength logging trucks traveled from the project site to mills in Redding and Weaverville each day, creating traffic hazards on the winding roads and highways. With a 25 foot overhang behind a logging truck, the end of the load is in the middle of the wrong lane on the right hand curve in the two lane road. It paid to take to the ditch in most cases when confronted with these loads of poles.

In Redding one day, I saw one truck with a load of 90 foot poles turn onto Court Street from the highway. The end of the load completely swept over the intersection just as a lady in a Volkswagen entered the crossing. She frantically drove onto the sidewalk, escaping being brushed off the road by the overlength poles. The truck driver kept right on going, oblivious to the near accident. But, not for long, I'll bet. That lady wheeled the V.W. around and took off after him with more than a steely glint in her eye. I'll bet that truck driver never forgot that lady when she caught him, which I'm sure she did.

While the clearing was in progress, an old time gold dredger was working the Trinity River just upstream from the mouth of Stewart's Fork. Homer, my fire prevention technician, had good rapport with everyone in the area, so he got permission for me to go onto the dredger and see the gold as it was worked from the riverbed. They were obviously in a rich area, as the volume of gold I saw made my eyes bug out. Even the dredger workers were excited. A few days later, though the dredger was sitting on the river bottom not far from where the rich strike was. Overnight, while the dredger was tightly tied to some trees, the river rose rapidly from a flash flood, and the water just went over the top. It was dismantled in place, thus ending an era of mining history on the Trinity River.

It was on the Trinity that I first met Henry. Now there's a man for you. A New Englander, crusty, outspoken, sharp, and hard not to listen to. He'd been a ranger, but was working on a special detail for the Forest Service related to the Trinity River project. He was, and is, an outdoorsman, who got visibly upset by bureaucratic red tape which interfered with production. Each day he was having to cross the Trinity River several times, and it wasn't always easy to find a safe or suitable crossing, so he bought a small, flat-bottomed skiff to expedite the job. It was O.K. to buy a boat, provided you went through the proper bureaucratic maze, which could take up to a month, but Henry just went to the store and bought a boat, without all the fol-de-rol of proper purchasing authorizations. "Henry, you can't just buy a boat," the administrative department tells him, a week or so after the christening of *Henry the Eighth*. They fumed and threatened old Henry for a while, while he glared and shouted back at them. Then it was decided he could keep *Henry VIII* provided he write a justification statement as to why he needed it and why he bought it. His statement was short and to the point:

"Only Jesus Christ could walk on water."

District boundaries were constantly changing on the Trinity side of the forest. In 1958 we acquired the North Fork drainage of the Trinity side of the forest. In 1958 we acquired the north fork drainage of the Trinity River. At the same time, Old Harry came to work for me as a crew foreman. He was old, lame, tough and determined. The last two adjectives more than compensated for the first two, and despite my misgivings about hiring him, he was a good foreman and a fine man; older and wiser than I. I didn't know he had heard me call him an "Old Codger" behind his back until he came to me one day in mid-season and asked, "Well, Bob, now do you think the "Old Codger" is going to make it through the year?"

Early in the season of 1958, Harry, Ray and I backpacked up into the Grizzly Creek country of the North Fork, just to learn the country. We built helispots, fished, hiked, explored and examined many of the old mines in the Trinity Alps wilderness. It was a week I'll never forget. Harry worked, hiked and did his share of everything. I knew by then that I had not made a mistake in hiring "Old Codger." By midsummer, his crewmen were well trained, and each was on his way to learning a trade from Harry. He was skilled in plumbing, carpentry, masonry and many other jobs. He had a personal goal of teaching each of his boys how to do at least one thing well during each summer season. He liked them and they liked him. Harry worked another two years before giving in to all those years and his lame back. A real man, is how I remember him.

Harry did have some funny ideas about rattlesnakes, which I can hardly believe, and the reason I bring this up is because of a snake incident which happened while working with him. We were working the Weaver Bally phone line one hot summer day and had just had a drink of cool water from a little branch of West Weaver Creek. I was standing beside the noisy little stream when Harry stepped up to me, put his hand on my shoulder, and said, "Don't move, and don't get excited, but you have a snake pinned down with your climbing spur." I looked down, and sure enough, there was about a two foot rattler pinned to the ground by the gaff on my climbing spur. He could only move his head about two inches and couldn't bite my boot. His rattles were whirring away, but couldn't be heard over the noise of the water. I remained in place while Harry "did him in" with a stick. When I cut the small rattles off, Harry warned me that the rattles would crumble and turn into a poisonous dust that could get in a man's eyes and cause blindness or be eaten by yellowjackets, causing their stings to become as lethal as a snakebite. I didn't argue with him. Who knows?

Weaverville was a small but historical community in 1956. Old piles of dredger tailings and eroded red hills surrounded, and in some places, were within the town. Even Main Street was paved only down the middle, with gravel and dirt between the road edge and the sidewalks. Locust trees grew in profusion along the curbs, creating a cool, shady atmosphere. Several stores had, and still have, outside spiral staircases from the sidewalks to the second story balcony, a striking and unique feature of the town. An old, but well kept bandstand, or gazebo still stands opposite the old jail, and the Chinese Joss House was not yet developed into the state park with guided tours that it is today. I suppose it was inevitable that it had to be taken into the state park system, as vandalism was beginning to become evident when we moved there. Pieces had been broken off various symbols and statues, and even boards had been ripped off the outside. The influx of people working on the Trinity River project no doubt hastened the rate of vandalism, and becoming a state park prevented further damage.

Scarred hillsides dot the entire area, and such names as Five Cent Gulch and Ten Cent Gulch probably indicated the frustrations of some old miners. A classic example of the devastation caused by mining activities is the old La Grange mine on top of Oregon Mountain, about four miles west of Weaverville, where the whole mountain top was sluiced away with giant monitor nozzles, using water from high mountain lakes some twenty to thirty miles away. A series of ditches, flumes, and siphons brought the water to the mine. It was an ingeneous system, even by today's engineering standards. A historical monument and parts of the old monitor are displayed at the site. Evidence of the siphon still stands out in the Stewart's Fork, Rush Creek area, where faint clearing lines through the forest and rugged slopes show up as they go up and down across the canyons. Active mining still continues by rugged individuals throughout Trinity County.

Deer and jackrabbits were often seen on the streets in town at night. One night at our house, right on Main Street, which is also Highway 299, I walked out to the garage to close the door. A light snow was falling, and as I walked into the building through the small side door, there was a clatter of hooves on the concrete as about six deer ran into the street. The next day, deer pellets on the floor proved to me that I had not been dreaming.

Another morning, during hunting season, Bernie, Dave, and I were just leaving town to go to work when a small forked horn buck stood in the street before we got out of town. He casually jumped over a fence into a yard nearby. That was a smart buck. He had jumped into Ray's, the game warden's yard. But, this trip had another deer story to follow. About three miles out of town there stood in the middle of the road about six does and fawns, plus a tiny buck with a spike on one side and a very small fork on the other. This little fork made him legal game, and I had my gun with me. Dave looked at me through his thick glasses, and Bernie grinned as I fidgeted while deciding whether or not to shoot that scrawny little old buck. "Surely you're not going to shoot him in front of his mother," Dave said, and Bernie added, "Not right in the middle of the road, Bob." "O.K., you wise guys, you run him into the brush, and in five minutes I'll go looking for him. If I find him I'm going to get him."

The whole bunch of deer reluctantly went into the woods as Dave and Bernie waved their arms and hollered. Five minutes later I started tracking them in the wet dirt, and sure enough, about 200 yards into the brush stood the little buck looking at me, still in the midst of the rest of the deer. Bang! I shot, and he dropped in his tracks, with a bullet in his head. As I walked up to him all I could see was the spike antler, with the other side buried in the dirt. The thought raced through my mind, I thought that was the forked side of the horns. Have I shot a spike by mistake? I gently, and with apprehension, lifted the head. Much to my relief, there was the fork on the other side. Both Dave and Bernie agreed that the deer they had chased into the brush had the fork on the opposite side, and I'm not sure yet whether I got the deer I was after.

Jim was my predecessor in fire control at Weaverville, and I had first met him some years earlier on a large Modoc fire, where he had impressed me as a good man and excellent sector boss. "He done good," was all that he wrote on my performance report, and he really wasn't a man of a few words. Due to some health problems he had been reassigned to a road maintenance job when I took his old fire control position. He and his wife, Cleon, became good friends with Betty and me, and consequently, I learned a lot from Jim and a lot about him. Though he hunted most everything, I think bear were his favorite game. It might have been that bear were essential to him because of his great love for hounds, which he always had in abundance. At the slightest provocation, Jim would take off on a bear hunt. He could and would follow them for miles and hours on end, and could hear their "music" for as far as their voices could carry across the canyons and hills. I got roped into a couple of his hunts, mostly by my own careless commitments. It happened innocently enough, like this:

One frosty morning Jim's son Dave and I were going out to Rush Creek to work when I noticed some very fresh bear tracks in the frost on top of an old buckskin log. "Dad'll be pretty upset with me if I don't tell him about these tracks," Dave tells me. So, I agree to drive back to town to tell Jim about the tracks that couldn't be more than an hour old. Jim was elated, and his elation was contagious as I agreed to go with him and Dave and the hounds. We walked, ran and stumbled uphill, downhill, sidehill, and everywhere following those dogs. "We'll have him up a tree in no time a'tall," Jim kept telling me, while his son Dave shook his head knowingly. Some six or so hours later we did have that bear treed in the biggest fir tree you ever saw, and the longest distance from the road you ever walked. It was a good thing we had an early start, 'cause it was pretty late evening when we got through dragging that carcass to the truck, and much later by the time we had it skinned. The hounds ate well that week. Jim always gave the bear paws to the old Chinese cook at the California Division of Forestry station, who prepared some exotic dish from them. He gave me the hide which I spent lots of hours on before losing the battle to maggots, flies and exhaustion. While working on the hide, Doug and his family from Salyer stopped by. His son Bruce asked, "What are you doing, Mr. Gray?" When I told him I was tanning a bear hide he said, "Looks like you're just wasting lots of salt, to me." I contemplated his seven year old wisdom, and took the hide to the dump.

The barn at Weaverville has long been gone. Its place is taken by an ultra modern high school; but the remembrances I have of it are clear—even vivid. First, it held tons and tons of hay. I know, I stacked it full several times, with runny nose, itching eyes, sneezing and sweating. Then I remember the good smells. Sweaty horses and mules—old creaky leather, saddles and harness well oiled and soaped by Joe, plus the hay and aroma of horse manure, old and new.

From the hay barn, a wooden incline, polished smooth from thousands of hay bales skidded down over the years, ended at the stalls on either side. Cats, mice and rattlesnakes also occupied the barns, though not enough cats because the cats would have eliminated the mice which provided the meals for the snakes. How many rattlesnakes? Well, there was nearly always at least one fresh dead one hanging on the fence to dry during the summer. Conservatively, I'd say one per week was killed. Believe me, much caution and poking around was done daily before dragging the bales down the ramp.

One day Betty was with me when I went to "feed the broncs." I was backing down the ramp dragging a bale with a hay hook, carrying a pitch fork in my other hand. On glancing over my shoulder I saw a large snake about five feet behind me at the foot of the incline. With a quick gasp and whirl of my body, I plunged the pitchfork toward the snake. One of the tines went through his neck (now, where does a snake's neck end?) about two inches behind his head, and pinned him to the floor, squirming and thrashing around. Why didn't that rascal rattle? I wondered until I saw that all his rattles were gone. I figured one of the mules had stepped on him and broken the rattles off. When I triumphantly brought him out to Betty dangling from the hayfork, she reacted as only my wife could when confronted by such horrible creatures as mice, yellowjackets, and rattlesnakes. She's a lovely, brave lady under all other circumstances.

It seems that I was always late getting home at Weaverville, usually because of the horses and mules. I had packed in supplies to Skip and Lev on the Cabin Peak Trail job, and it was well after dark when my horse, two mules and I reached the north fork of the Trinity River. The water was a little high and we'd had some trouble crossing in the daylight, but I decided to try anyway. With much reluctance, "King" entered the water. Soon I was lifting my feet to keep them dry, as water rose up past and above "King's" belly. After what seemed like an eternity we reached dry land and started down the trail parallel the river. After about five minutes, I thought to myself, what's the water doing flowing that direction? I was obviously traveling upstream and knew I should have been going downstream. "Whoa, you sons of guns, we've got to figure this out." After dismounting I realized what had happened. Instead of crossing the river, "King" had waded downstream and finally turned back to the same bank we entered the stream from, heading right back to where we'd come from.

I unsaddled, fed the bronc and mules, and bedded down for the night, making a leisurely trip to Weaverville the next morning. Even in daylight, the crossing wasn't too easy.

ALL ABOUT MULES

"A mule is an animal with long funny ears," goes one of the verses of an old song about monkeys, pigs and other critters. Besides this unique characteristic, mules are a lot more. Size, shape, character, color, all different, and each with his or her own personality. With a mare for a mother, a jack-ass for a father and unable to reproduce, he (or is it she? or it?) vents its frustration in various mulish ways.

First, it was Cactus in McCloud, then Joe in Weaverville who watched me really learn about these interesting animals. Now, I had some experience, or you might say, exposure to mules when I was going to college. We had some livestock judging classes in which I did very well, except for the mules. In groups of beef cows, milk cows, hogs, sheep and horses, the professor and I had good rapport and we nearly always agreed on the ratings, except for mules. My number one mule was always his number four, and with the remaining three as mixed up as possible in our judgment. With our differences of opinion, plus him overhearing me muttering under my breath about his stubborness regarding the virtues of these animals, and their similarity to him, my strong A was reduced to a B- in Judging Animals. It probably would have dropped to a D, except that he was one of the supporters of the annual L.S.U. student rodeo, and with his urging, I agreed to enter the bareback mule riding event. His number one choice threw me after about two leaps out of the chute.

We'd better get back to Forest Service mules. They're much smaller than the draft type plow mules I'd seen all my young life in Louisiana, and much better adapted to the packing chores for which they were used on the Shasta-Trinity Forest up until the past few years.

Packing had been an occasional job at McCloud and Fall River Mills, where we never had more than two pack animals, but my arrival on the Weaverville District introduced me to

And Here's A Mule

some eight or ten of the long-eared beasts in addition to the saddle stock. Joe, the regular packer had an ailing back when I arrived, so I inherited the packing job along with my fire control duties.

Dexter was the oldest, biggest and most dependable of the string. He had a big, low slung belly, and was easy to work with.

Charlie was tall and rangy, with a good disposition, and would only kick you when he had a chance. With an easy gait, he was also a good saddle mule, looking like Festus' mount in the old "Gunsmoke" series.

Doc had class, but whenever I had mule troubles, he was in the midst of it, like falling down in the stock truck or kicking through the racks for no apparent reason. It took lots of patience and understanding to work with Doc.

Job had the patience to go with his name, and was the only one I ever knew that would let you grab onto his tail to assist in going uphill. Everyone liked Job. Lois, Flossie, and Nevada, all mare mules. Dainty, delicate and dangerous. One would nuzzle you while the other kicked you. They were easy to load and fun to use, except for Nevada. She had everyone buffaloed. Skittish as a sackful of cats, and had a big scar on her rump. Joe said the scar was from a mountain lion when she was a foal. I don't know that for certain, but it made a good tale and an excuse for everyone not to use her. Well, I decided to calm her down and pack her. It was remarkable that once she was saddled and loaded, she turned into a gentle, cool, and calm lady who could pack her share. I think she was smart enough to know she wouldn't have to work if she put on her show each time we used the mules, because she continued to be nervous and unsettled until the pack saddle was in place.

Granite Peak was a long day's trip from Weaverville. Loading the stock about daylight, driving for two hours in the stock truck to the trailhead, unloading the truck and saddling and packing the broncs usually lasted until 10 a.m. It was a steep four mile climb, and in the springtime, much of the trail was covered with hard packed snow, which necessitated digging tracks for the mules to walk in across the steep barren snow covered slopes.

On one occasion, before learning the need to dig steps, one of my two mules slipped on the rock hard snow and slid about 100 feet down the hill. Sheer luck resulted in him sliding into a soft brush patch rather than into one of the huge granite boulders. Except for a few hairs rubbed off and the load being scattered, Doc was all right. Dexter, from the top of the snowback, looked in contempt at the mule who lost his footing and patiently waited the hour it took Doc and me to recoup and regroup.

I later found out that another mule had been killed a few years earlier at this same location under the same snow conditions.

I seldom went anywhere with the packstring without some misadventure. Let me tell you about the most memorable one, which was my own fault for not using the best judgment.

It was late summer, the trail crew had finished its work in the Stuart's Fork drainage, and I took my horse, King, with the seven mules to Morris Meadow to pick up camp and haul out garbage.

The day was beautiful, the mules were at their best, possessing most of the virtues of the Boy Scout Law, excepting "thrifty" and "reverent." Ten out of twelve ain't bad in anybody's book.

We got to the meadow in midafternoon, with lots of daylight left. The trail crew had already left, and I'd planned to camp overnight, leaving in the early morning. Well, I got to thinking, I'm not tired, the mules are already saddled, and with a little luck, I can make it back to the Trinity Alps Resort shortly after dark.

The mules' dispositions started to change as I loaded them, and even King looked at me in a strange way. No doubt they had anticipated a good roll in the meadow and an evening of foundering themselves on the tall grasses which they could eat without stretching their necks. With about two hours of daylight left, we headed down the trail. In my hurrying, I didn't balance the packs as well as I should have, and within a mile I was already adjusting, balancing and retying my self-designed diamond hitches. With darkness gaining on us, we settled into a slow trot, with packs bouncing and lead ropes straining between the animals.

My lack of wisdom in trying to do two day's work in one was rapidly becoming more apparent to me as we rounded a bend, coming face to face with a bear. I reined up, and the mules piled up behind me on the narrow trail. The lead mule was probably the only one to see the bear, but that was enough. He whirled around, pulling the lead rope from my hand, and crashed into the mule behind him. Like falling dominoes, the reaction took place. Each mule whirled into the next one, and in seconds, lead ropes were broken, packs were flying, saddles hanging under bellies, and mules scattering into the trees and up the trail. Fortunately, it took more than a bear to get King upset, so I dismounted and tied him to a tree as the bear disappeared up the hill.

When I walked back up the trail to survey the damage, I was appalled to find packs, saddles and broken harness scattered for a quarter of a mile up the trail and some of the mules disappeared into the woods.

Darkness was rapidly approaching, and I couldn't find a flashlight, so I had better take care of the animals first and worry about the gear later, I thought. Catching the mules wasn't too difficult, and as I caught them, I removed what was left of their saddles and tied them to the nearest tree. By the time the first four were taken care of, it was dark and all I could see were dark shadows as I approached the others. Never being sure which end of the creature I was approaching, I talked gently and soothingly to them, so that if it was the kicking end, maybe they'd kick gently and soothingly.

Finally, I had found and tied up six of the seven, but could not find the seventh. "He'll just have to wait until daylight," I said to myself, hoping that he was not tangled up in the brush somewhere.

Finding a sleeping bag and something to eat was not too

difficult, as I remembered where they were laying along the trail before darkness came. A cold can of beans and some crackers sufficed before crawling into the sleeping bag for the night, but I didn't sleep too well wondering where the seventh mule was.

Daylight came and I shivered as I came out of the warm bag. "Gotta find that mule," I muttered to no one in particular. "One,—two, wonder which is missing,—three, four,—five,—six, seven!" They were all there. Somehow I'd forgotten to count the first one I tied up after the bear. "Darn, I sure could have slept better if I'd known all were accounted for."

A half sack of grain was distributed to the eight animals, and they "ate like horses," if I may make a joke. The grain didn't fill their empty stomachs, but helped until I could get the saddles patched up and loaded, which took about four hours.

I was tired and the animals hungry, and the packs pretty scraggly looking when we arrived at Trinity Alps about two hours later. A couple of bales of hay and a late breakfast furnished by the Resort restored the mules, the horse, and my spirits.

When I arrived at the barn at Weaverville, Joe and Bernie were there to help me unload, bless their souls. "How did it go?" they asked, "You made good time," not yet knowing what I'd been through. "Had a good trip," I lied. "Oh yeah, then what the hell happened to this pack gear?" asked Joe, who was going to have a few days' repair job ahead of him. Then I told them the story. "It couldn't have happened to a nicer guy," they agreed. A couple of sadists, I thought, as they obviously enjoyed the story of my trauma of the past 24 hours.

Moving to Mt. Shasta took me from all the mulish problems of the Trinities, to the horsey problem of Buck. Packing at Mt. Shasta consisted only of an every other week trip to Black Butte Lookout, which was an easy pack of only two and a half miles. Tony was a quarter-of-a-century old horse, with personality, pride, dignity and gentleness, and we used him as a saddle horse. My horse King was also gentle; one whose disposition would allow two men to shoe him, a front foot and the opposite hind foot, at the same time while he dozed.

Then there was Buck, flighty, unpredictable and I think, somewhat deficient between the ears. Anything, or nothing at all would send him into orbit. Whether true or not I don't really know, but the rumor was that a beating across his head with a chain some few years earlier had been the cause of his unpredictable behavior. As with other problem employees in years past, I accepted Buck as a challenge, and usually when I worked alone with him we got along quite well. Bob H., who did much of the packing, and Buck had a mutual understanding. Each disliked the other with undisguised distaste, so Buck always put on his best display for Bob. Larry M., too, had a distrust for Buck, but could always see a little humor in the situations between them.

"There, you son-of-a-gun, you're all packed," said Bob as he completed the last knot on the pack. I had my camera ready to take a picture of the beginning of our trip to Black Butte when Bob made his utterance. Buck lived up to his name and reputation; five gallon water cans flew through space as Bob scrambled aside. Acting on instinct, I snapped a picture. One can was in mid air and another resting on King's back, neatly between his already tied down packs. Buck was hunched over and dust flying, as the picture later showed. A half-hour later, we reloaded Buck and went on to the lookout without further ado.

Larry M. and I were involved in the next Buck episode.

An innerspring mattress is not the most handy item to pack, so we loaded it on King with the corners bent down and tied. Buck was following King as I was leading him on foot. Larry was a respectable distance behind Buck, also on foot.

Larry called, "Bob, hold up, your mattress is coming loose." I stopped, and Buck must have looked back just as a corner of the mattress sprung out from its rope. This really offended Buck's senses. He charged at me, as I dived off the trail. A few feet further, he also went headfirst over the side with packs flying off and Flamo tanks bouncing through the rocks. He tumbled down the steep rocky slope, winding up with four feet in the air wedged against a tree. Poor old Buck, that's the end of him, I thought. But no, when we reached him, he was wide-eyed and bushy-tailed, wondering how to get up. With our help he succeeded in getting to his feet. There was lots of bare horsehide showing, but apparently nothing broken. We reloaded him and Larry took off up the hill with the two horses. I made some lame excuse about having to get back to town early. Do you suppose I "chickened" out on Larry? Never have felt good about that incident.

One of the horsey problems at Mt. Shasta was with old King, my own pinto, who stayed at Kaiser Meadows or in the lot back of the Forest Service office. He had several bad habits, mostly to do with taking leave from his pasture by squeezing through openings far too small for his generous bulk and heading for Manuel's garden at the nearby railroad section house. Sometimes he'd go to Mrs. Barnett's petunia patch, and other times to Mrs. Schmidt's pasture with other horses. He'd get into her pasture by walking over the cattle guard, carefully placing his big feet on the bars as he walked across.

Another trait that he possessed was an aversion to the barn. In rain, snow, blizzard or cold, he preferred to remain outside, pawing through the snow to reach some frozen, dead grass or weeds in preference to fresh dry hay in the comparatively warm barn. He also like to roll in mud, which did nothing for his looks with the shaggy winter coat.

Well, it seems that Mrs. K. of the Humane Society saw King looking his worst while pawing through a foot of snow for whatever he could find to eat. She called me to take proper care of my starving horse, not realizing he was choosing his way of life and shunning the racks full of hay in the readily available barn. Neither could she see that he was about 100 pounds overweight.

Wow, I thought to myself, I've never been accused of

being inhumane to anything in my life, so I made sure that my horse was not being neglected. Sure enough, he was full, fat and happy, despite the miserable weather going on around him. I quit worrying and kept the barn door open and the hay racks full, thinking Mrs. K. had accepted my explanation. But no, a few days later the phone rang again—"Mr. Gray, if you don't do something about that horse, I'm having you arrested." Well, I got a little irritated this time and told her that King was all right, and suggested she drive out Old Stage Road and look at several other horses and a herd of Angus steers also out in the fields with their rumps turned toward the blizzard with no barn to go into even if they wanted to. She didn't have me arrested, but continued throughout the winter to give me a pretty bad time whenever the snow got deep. Which brings me to the climax of my dealings with Mrs. K. . .

My telephone rang one snowy spring day, and the receptionist told me "Bob, Mrs. K. is on the phone." I answered an excited, high pitched voice. "Mr. Gray, Why do you have your dog tied in the back of your pickup in this weather?"

Sure enough, there was a big black Labrador happily sitting in the back of my pickup across the street. It dawned on me that my neighbor's dog had followed me to work and jumped into my pickup when I went into the office. I blew my top, telling her, "That's not my dog, and he's not tied in my pickup!" as I slammed the receiver down and hurried down the hall toward the coffee pot, thinking very unkind thoughts of Mrs. K. Then I met John coming down the hall with a big smile on his face. "What's the matter, Bob,—Mrs. K. giving you a bad time?" Then it came to me. I'd been talking to John all this time and he fooled me completely.

And, you know, I never heard from Mrs. K. again!

MINING—SURFACE RIGHTS

The forests of Northern California bear remains and scars from the days of the forty-niners, and the following years of active gold mining. Picturesque and sometimes descriptive names of mines dot the maps of the Shasta-Trinity and Klamath national forests. Women's names were favorite choices for the names of the mines; such ones as "Ruby Pearl," "Little Anna," "Mary Blaine," "Milkmaid" and "Hazel D." come to mind. Men's names are not totally left out either with "Crazy Jim," "Bob's Farm," "Long Tom," "Hiawatha" and "Lone Jack," to name a few. Geographic names, probably from the miner's younger days, have a good showing. "Alaska," "Hoboken," "Ozark," "Altoona" and "Uncle Sam," along with "Woodrat," "Brown Bear," "Mule Shoe," "Mad Ox," and "Mad Mule," representing the animals. Some just had good sounding names like "Old American," "Star City," "Gypsy Queen," "Bonanza King," "Last Chance," "Boomer," "Kena," "Bully Choop," "Lady Slipper" and hosts of others with numerals added when original names ran out, as "Luck Strike # 1, #2, #3, #4, and #5."

Well, along about 1958, the federal government decided to determine the validity of these myriad claims, so a program of Surface Rights Determination was originated. Hundreds of man-days were spent by Forest Service men walking trails and driving remote roads and Jeep trails, searching out evidence of work being currently done or having been done within the past three years. Some claims had cabins on them, most of which were crumbling into various stages of disrepair. One out of ten might be occupied or show signs of intermittent use, and maybe even have some fresh "diggins" nearby, all of which was duly reported on the forms provided. It was a fun project in its infancy, provided you liked hiking and wandering the back country. On a minor side trail off the north fork of the Trinity River, I was examining a mining claim when I suddenly felt someone's presence nearby. I was standing in a small open meadow and turned slowly to see who or what had caused my feeling. There on a log about thirty feet away sat an unsmiling old man with a slouch hat shading his face. A rifle was casually resting across his legs, but also casually pointed in my direction. I know I jumped when I saw him, because I never expected to see anyone in the remote area. After regaining a little of my shook-up composure, I introduced myself, to which he did not reply. "Is this your claim?" I asked. He nodded his head affirmatively. Then I volunteered to him all the information about the Forest Service Surface Rights Program. He listened, but was unimpressed. When I headed back down the trail some fifteen minutes later, he was still in the same position, and the rifle was still casually pointing my way. The few seconds it took me to get out of his sight seemed like an eternity as I walked away with my back to him. When I got out of his sight, I broke into a fast lope, headed for the pickup. It really was getting late in the day.

Within the next two or three years, owners of claims declared invalid were notified and told that any cabins or structures on their claims would be torn down or burned by the Forest Service. Some protests were made by owners who had nice cabins, and who felt they had legitimate claims,— "one on which a prudent man could make a reasonable profit from mining." Admittedly, some so-called claims were no more than a free homesite on government land; but times were changing and this situation was no longer acceptable to the U.S. government. I personally was involved in the burning of two cabins, and it left me with a very unsavory feeling. Though they were not elaborate, nor even in good condition, they seemed to me a part of a fading era and I was hastening the end of that era. Afterward, I chose not to participate in burning old mining cabins.

It was still a thrill to see someone really taking some gold from the mined-over countryside. One day in Canyon Creek, Mr. K., a miner, was doing his thing with a backhoe and large sluice box. Each bucket of gravel produced large nuggets as he ran it through the riffles in the sluice box. He said it was only a "pocket" and not a vein, but would produce gold about "back to here," as he pointed to a spot on the hillside. I'm not very emotional about gold, but got kind of a tingle of gold fever as I watched him work.

MEMORABLE CHARACTERS

Because of the Forest Service, I've met a wonderful, interesting and remarkable aggregate of people, who might be called clients, affiliates, associates or just characters. Among them are ranchers, miners, loggers, housewives or any one of varied classifications.

One of these was Amos. His shanty up the North Fork from Helena clung to the hillside above the beautiful stream. For a year or so I had seen him at various places along the road and highway between Helena and Weaverville, always with hip boots folded down around his ankles. He walked vigorously with a jerky motion and, as I later found out, talked the same way. After passing him by for a long time, my conscience began to hurt, so I picked him up. He was a miner, and unlike many in Trinity County, he had no ill feelings for the Forest Service. In fact, I'm not really sure he was even aware of us. A simple man with a simple lifestyle, but happy to have a friend to talk to and who would give him a lift or bring him something from town. His coffee was "out of this world," the few times I could find no reasonable excuse to decline. Just one of the hazards of the job, I decided, as I downed the coffee. Where he mined, I could never determine, but his house was full of five gallon buckets of black sand, or heavy sand, as he called it. It was all a man could do to lift or move one of the buckets. Not being a mining authority, and his tight lipped air of secrecy prevented me from ever learning what it was. Nevertheless, he liked showing me how much he had gleaned from his source since last visiting him. Once he told me he made lots of money, and even though his lifestyle didn't show it, I believe he did. Other miners in the area said he was a little off-center, but, I never saw any "heavy sand" at their homes.

Judge M. was the justice of the peace at Junction City. A man very critical of the Forest Service, especially concerning "light burning." His wooded hillside above his home bore out his thinking because, indeed, it was a beautiful park-like area with no underbrush or litter. Each spring he let fires creep through the needles on his property, doing a thorough job of hazard reduction just as we of the Forest Service encourage home owners to do. His thoughts were good, but unrealistic from our standpoint, because of cost per acre and risks involved. Later on I intend to go into this subject more thoroughly, as many people both within and outside the Forest Service see this as a cure-all for fire control problems.

Though our philosophy was different, I liked the judge, and determined to stay on friendly terms with him, and also to continue my job in law enforcement within his area of jurisdiction. In fact, he was a challenge to me, so I made certain that I would get involved in all cases he handled regarding fire law violations.

The first case I brought before him was a licensed packer up Canyon Creek. Seems that one late afternoon his pack string stirred up a yellowjacket nest along the trail. In his frustration at having his packstring scattered and himself being stung several times, he set the yellowjacket nest afire. Well, I sometimes admire a man who takes charge and does something, even though it's wrong. He put the fire out, he thought, but a hiker coming down from the high country late at night discovered a small fire along the trail and at midnight he reached Weaverville, where he reported it to me. He had stopped its spread and said it was only about 10 feet in diameter, so I decided to go check it and finish putting it out by myself. On reaching the fire, about two miles up the trail, and several yellowjacket stings later, I had the fire out cold. A little detective work, like checking footprints of milling mules, a few horse apples, a chunk of lash rope, and some gear scattered in the brush, correctly led me to believe a packer, probably from the trail head corral had something to do with the fire. Sure enough, about daylight I stopped by the pack station in time to talk to the packer. "Yes," he answered to my question, "I burned out them damn yellowjackets. Every time I go by there they sting me and my mules."

I had everything for my case. An admission from the packer that he burned out the yellowjacket nest, a statement from the hiker that he discovered and worked on the fire, my own statement, and four or five welts from yellowjacket stings on my anatomy attesting to my work on the fire. Armed with all this, I filed a complaint at Judge M.'s court.

On the given date to appear we all showed up, the packer, the judge and I. The packer plead "not guilty," and the judge confirmed his plea. I got a lecture on the beneficial effect of fire on the woods, including, "The Indians burned the forests, and we didn't have all this underbrush that the Forest Service has allowed to grow." The admonishment to the packer was, "Now, next time be sure to put your fire out, we have to get along with the Forest Service."

I was flabbergasted, frustrated, and mixed up. Who was on trial, I thought, me, or the packer? Determined not to let this get to me, I asked the judge if I could talk to him after the trial. He then said the reason I lost was that the packer had done what any prudent man would have done in similar circumstances. Scratching my head, I left his chambers. Next time things will be different, I mistakenly thought to myself. The next time wasn't long in coming. Mr. B., a small rancher up Connor Creek had a rather large burning job to do. He and I took about two hours going over the requirements on the burning permit. A day or two later, on checking his burning operation, I found a few details he was not adhering to. For instance, he didn't have adequate lines around his fires, they were not always attended during daylight hours, and he paid little attention to the times prescribed to light new piles. In general, our two hour discussion when preparing the permit was just so much smoke. I was pretty upset with his lack of compliance and poor attitude, telling him, "If you don't have all the permit requirements met in 24 hours, I'm lifting your permit and we'll visit the judge. Not only that, but we'll put the fire out and send you the bill." He seemed rather unconcerned as he grumbled some reply while I was departing.

Within 12 hours I had been dispatched to a fire on the Angeles Forest, and had to tell Bernie, the prevention man, of my threat to Mr. B. Bernie was to check on the conditions the next morning and carry out the issuance of a citation, if necessary.

When Bernie arrived the next morning, Mr. B. was nowhere to be seen, the fire had escaped and was on its way up the hill. About fifteen acres burned before the Forest Service brought it under control, and Mr. B. had still not showed up. News via the grapevine travels fast, though not always accurately. Before the day was over, word got to me on the fireline of one of the many Angeles Forest fires I've been on. Thankfully, the report I received was worse than the facts, as I was told several hundred acres had burned. They had just as well sent me home from the Angeles then, as I was not much good to them the rest of the shift. A telephone call to Weaverville, put me more at ease, when I finally got the facts from Bernie. "Get everything on paper, Bernie," I said, "and we'll at least 'hang' the guy in court." (Not a literal translation, you know, Mr. B.)

A week later, back home from the Angeles, we prepared our case. An airtight prosecution, with all the details on paper, plus the assistance of Don, the Trinity County D.A. practically assured us of a conviction. Judge M. told me he would have to disqualify himself, as he was a friend of the defendant. This should really clinch our case.

At the appointed time, Bernie, the D.A. and I arrived at the judge's chamber. There sat Judge M. He had not disqualified himself. Neither had he told Mr. B. about the trial date, so he had to make a telephone call to his home in Redding. His conversation went something like this:

"Hello, Bill, this is Hank. It seems you're in some kind of trouble with the Forest Service. Something about a fire—When can you come over to get it straightened out?—O.K., I'll see you on _____."

Don shook his head in disbelief. My determination to be friendly with the judge was severely undermined. Well, on the new date, we met again, this time with Mr. B.

My diary, dated December 14, 1959, reads as follows: 12:30 p.m. To Junction City Court. Took Scott, Kennedy, Leas and McCaffrey. Presented our case. Mr. B. presented his side. Judge M. heard only one side of the story.

"Not guilty," sez the judge. Guess he didn't hear Mr. B. say he lighted the fire that went up the hill. They should have this case on T.V., it really sets a precedent.

Final comment, "I'll be damned!" With the final comment, Judge M. pointed his old crooked finger at me and said, "If you are interested in a contempt of court citation, one more word from you will do it."

I remained quiet as Don, the D.A. applied a little pressure to my shoulder.

My goal set in 1956, to develop a good relationship between the judge and the Forest Service had failed, as I transferred to the Sacramento District a few weeks after this trial. This failure has bothered me ever since, as I feel part of my job was to bridge gaps of misunderstanding between our agency and the public.

It was a damp, early April day when I saw the tall column of smoke. Better see what's going on, I thought, as I drove to the brushy hillside. On approaching, there was an ancient looking bulldozer with a plaid-jacketed operator who was clearing and burning brush. I watched a few minutes as he deftly handled the old machine, pushing brush onto already burning piles. A nice job, I thought, as I walked up to the dozer. Mr. D. looked at me from the tractor with the motor idling. No way could I talk to him over the noise, so I beckoned him to come talk to me. Now, this guy's got a chip on his shoulder, I'm thinking, and you know what? I'm right. "What do you want?" he asked, pretty gruffly.

"I'm Bob Gray, with the Forest Service, and new to this district, so thought I'd check your smoke and introduce myself." After some favorable comment on his clearing operation, I asked if he had a burning permit. "No," he answered. "Did you know you need one this time of year?" "Yes," was his response. "Well, why didn't you get one before burning?" I asked a little bit impatiently.

"Every year I ask for one," he replied, "and every year I get the same run-around, 'too windy, too dry, too late in the season,' from you guys." He really opened up to me. "You

176 / Forests—Fires & Wild Things

know, I've got a job to do and have to do it whether the Forest Service approves or not, and I knew I couldn't get the permit, so I just went ahead with the burning job."

"Mr. D., you don't leave me much choice but to cite you into court. If you want me to, I'll radio in for the time to appear." "The sooner the better," he replied.

The judge could see him within the hour, so I asked if he'd like to ride to town with me. "Yes, I would. You know, I like you." This startled me as he went on saying, "I've been arrested before by the Forest Service and all the other times they sent the sheriff for me."

We had an interesting talk about his ranch, the clearing job, and the Forest Service, as we drove to town and back to his ranch after the court appearance. Yes, he lost the case with a suspended sentence. I gained a friend, and even his ill feelings for the Forest Service were improved.

A year later when his grandchildren started a brush fire, we had an enjoyable visit and he complimented our efficiency in suppressing the fire.

After fire season in about 1947, I helped Jess maintain the McCloud Railroad firebreak. Both of us were bachelors, and lived at the barracks in McCloud, cooking for ourselves. I hated to cook and Jess liked to cook, but couldn't. Fortunately for me, I was squiring my future bride and had a good supper at her house each night, so only had to struggle through breakfast and lunch with Jess. Now, breakfast wasn't too bad but lunch was the same, leftover fried eggs and cold soggy hotcakes; consequently I ate big suppers and not much else. Hotcake and fried egg sandwiches still don't do much for me.

Jess was a delight to work with. Witty, funny, hard working and a little profane, he had been a cowboy during his youth, and his legs looked like a couple of parentheses. This, plus some poorly mended broken feet made him walk with a distinct rolling motion. No complaints from Jess, though, as he went about his work running the bulldozer, while I did the swamping for him. This included throwing a few rocks, cutting a little brush, fueling the tractor, and keeping the warming fire going. On occasion I'd operate the dozer for short periods, and once I really messed things up. It was a rainy Monday, and Jess had come to work with a new "tin suit" (rain clothes) and boots. The rain stopped about noon, and I got on the Cat to give Jess a rest. The first thing I did was to back the tractor onto a boulder, and spin the rig on its tracks, not knowing that Jess had just laid his rain suit on the same rock. Then I saw Jess waving frantically and hopping around in his sock feet. I still didn't know I'd demolished his new tin suit, but when he pointed to the rock I saw what I'd done. He then threw the new boots at the rock. "Go ahead, grind those up too, you 'blankety-blank idiot.' " I felt pretty bad, but soon Jess was laughing and joking as usual, saying, "Guess I'm just lucky I wasn't still in the suit."

He reluctantly accepted the new outfit I bought for him that afternoon. Whenever I operated the dozer after that, Jess made much ado of being well away from me, usually peering from behind a large tree or rock.

PORCUPINES & SUCH

"You throw away the porcupine and eat the plank." So ends the recipe for "planked porcupine."

"You don't kill porcupines in the forest, because it's the only animal a lost, unarmed person can kill with his bare hands and a stick." (The stick is most important.) This may be true but I've yet to talk to the person who has tried it, and I'm an experienced finder of lost persons. Here's what I know first hand about the prickly little creatures.

They make funny tracks in the dust, like dragging a broom behind a beaver. They're not very pretty, not very smart, don't look like they're enjoying life, and I'm not sure their mother even loves them. They're too slow to get away from dogs, which aren't smart enough to let them get away, and they don't show up well enough in headlights to keep from getting hit. You don't know why your tire is flat until you take it off and find a porcupine quill in it.

What do they eat? Your saddle, your boot, your belt and the sweatband in your hat. Then your rifle stock, saddle blanket and other boot. From there it's the front doorstep and the bacon.

When deprived of these delicacies out in the wilds, it's the plywood sign you placed last summer and the routed sign with linseed oil.

But their favorite food is the 100 acres of pine seedlings planted last year, or the bottom foot of the five foot trees in another plantation. Sometimes an acre of natural reproduction is their choice, where they girdle all the trees at nose height. Creeping Destruction, they are.

Dutch and I were driving home from Medicine Lake one night when we saw a lumbering shape in the road. He must have been the granddaddy of all porcupines. "There's a shovel in the back, let's get him," says Dutch as he came to a screeching halt. I grabbed the shovel and approached Old Grandad from the side, failing to see a round rock about baseball size. As I swung the shovel my foot hit the rock. It rolled and I fell right over that porkey's back, with the shovel flying. My body made an arch over him as he ambled on into the darkness. "Did you get him?" Dutch called from the cab. Well, no, I didn't get him, but I felt pretty lucky as I picked a dozen quills from the front of my shirt in the headlights of the pickup. Dutch just grinned and relit his pipe.

On another occasion I let one get away. We were looking for a lightning fire in the McCloud Flats and couldn't find it, though we knew it was nearby. I climbed a pine tree about 80 feet tall to get a look around. At about 50 feet I met a porcupine face to face two feet in front of me. He hardly even bristled as I detoured around him on the way up. Coming down, he was still perched on the same limb, with his dark beady eyes fixed on mine. We gave each other the once over for about ten minutes, just seeing what the other species looked like at close range, until my crew hollered, "What are you doing up there?" "Just communicating with a porcupine," I answered, "Oh, yes, the fire's about 100 yards that way," and I pointed to the fire. The crew hadn't seen the porcupine until I got down and showed it to them through the dense branches. "It could only happen to you," said one of the crewman. Oh yes, it did happen again while deer hunting with my father-in-law. He's a tree hunter, and had coaxed me up a tree while he ran a brush patch. I let him go too.

Ringtails are meek little catlike animals with a tail like a racoon, but more delicate bodies and somewhat sharper features. There's a family of them at Black Butte, or was when Ann was the lookout there. They are night creatures, as Ann discovered one night when she was outside the lookout with her flashlight. She decided to trap them, in which she succeeded admirably, capturing as many as seven in one night with a figure four trap. "They're not the sharpest animal in the world," she said, "because some of them I'm caught twice in one night."

On one trip to the lookout, she gave Ike and me a ringtail to bring to town. We put it on Ike's screened porch for the night, and you know what? That creature disappeared. We don't know how it escaped, because there were no holes in the porch.

A short time later there was a write-up in the *California Outdoorsman,* telling all about ringtails and their habitat. Black Butte Lookout was definitely not a likely home for them, but they were there. You can ask Ann.

Let me tell you something else about Ann. It seems the year before I came to the Sacramento District, Ann cut her leg with an ax while at the lookout. Stitches were required, so she sewed it up. Now that's independence, resourcefulness and individuality wrapped up in one *character!*

Not too many people have seen a wolverine, unless it was in a Disney film or a zoo, but I got a real look at one close up. It was on the north fork of Trinity River, near Strunce Cabin. My horse, King, and I were resting along the trail in an open flat with large virgin Douglas fir bordering the North Fork. King perked up his ears and looked to his left and we saw a brown, low slung, furry animal bounding toward us about 75 feet away. He was running as though something was after him and passed about 20 feet from us in plain view, but evidently wasn't aware of our presence. About 100 feet from us was a small heavily leaning tree about 20 inches in diameter which he scurried up to a height of about eight feet. From here he jumped back to the ground and ran right back by us. King never stirred as we watched his erratic antics. On getting back to town I naturally read up on wolverines. The book did show the North Fork area to be populated by the rare bearlike animals. In talking to game biologists later on, they told me the erratic behavior was quite common with them, and they are pretty vicious with other animals.

Marmots, they tell me, are kin to the groundhog, found at Punxsutawney, Pennsylvania, and other points in the eastern United States. Maybe we should have a Marmot Day in the West to determine the length of the remaining winter, because we do have marmots out here. More specifically, on Granite Peak, and they live in burrows, like their eastern counterpart. It so happens that the colony on Granite Peak elected to build their burrows right in the middle of the trail to the lookout and one of my loaded mules fell through. Both hind legs were in up to the hilt, if mules have hilts. I think Rod, the Minerville crew foreman, was with me. Anyway, we tugged, dug, and shouted in his ear to get that mule out. He just lay there. Finally we unloaded the pack, and pushed dirt and rocks in the hole under his belly. When we got enough in there, he climbed out unassisted. In all my trips to Granite Peak, though, I never saw one marmot.

One seldom sees a beaver in the wild, but their handiwork shows. I've never really doubted stories of their industrious nature, but am somewhat skeptical of a few tales about them. For instance, as tree fallers they leave a lot to be desired. If a tree falls where they want it to, it is sheer luck. As you look around a beaver dam, you'll find trees felled uphill and away from the stream, trees partially cut off and hanging up in other trees, and trees gnawed halfway through and abandoned. Occasionally, one is felled right where it belongs. But their dams! Now, that's a different story. They are really crafted by experts. The mixture of limbs, twigs, grasses and mud is unbelievably intricate, and a puzzle to unravel. Most dams are a beautiful arc between two higher points of land at each end of the dam, and sometimes reach over 150 feet in length, forming a quiet pond behind, with willows and other water tolerant plants growing in and around the edge. With this background of the typical beaver dam, let me tell you of the one I contended with.

In 1945, early spring as I recall, Cactus, Don, the game warden and I had six young beaver in a large cage, and three small empty cages. Don had been told by the California Fish and Game people that there were three males and three females. We were to match them up in pairs and plant them at three pre-selected locations on the McCloud District. We were totally flabbergasted in trying to determine the "he or she" status, and in fact all looked alike to us. We decided to put two in a cage, and if they didn't fight with each other, it was a good matchup. Evidently our method worked, because after a few years, beaver colonies were established on Trout Creek, the McCloud River and Dairy Creek, a tributary of Squaw Valley Creek.

Some 14 years later, my wife and I bought a few acres of land on Dairy Creek, and some signs of beaver activity were evident along the stream, but no sign of dam building. A couple years after that the McCloud Lumber Company built a logging road alongside of the property and installed a seven foot culvert about 50 feet long where Dairy Creek crossed the road. The superintendent of the beaver colony, dam building division, decided the culvert head was the ideal place for a dam. He wanted something different than the run-of-the-mill beaver dam, and the logging road fill with a few sticks strategically jammed into the culvert would provide about a ten acre lake having everything a beaver could ever want. And look at all the work it would save.

For a few years the lumber company kept tearing out the dams, as the water backing up would overflow their road and wash out the fill, but as their need for the road ended, the dam destruction job fell to my uncle M., my father-in-law and me. Pretty regularly we would take it out, and just as regularly they would put it back. The secret is to do it before they get it completed, because when complete, it nearly fills the culvert head and slopes into the pipe about 15 feet, making a tough job, and a little bit dangerous as the removal on a March day in 1977 proved.

On this particular day, some 32 years after planting the original pair, my wife and I went down to remove the dam, which this time was about five feet high with the crest of the dam about five feet into the culvert and a real quality job of construction. Every limb, twig and stick was firmly embedded in long grasses and mud. I had gone up inside the pipe from the outlet end and pulled all the loose sticks from the lower side of the dam, letting them float through the pipe in the shallow stream formed from water going over the top of the dam. Betty was a little worried as I opened one end of the dam, and the water started rushing through. I kept pulling and prying out sticks as the cold water soaked my legs up to my knees. After getting one side running and eroding the mud away, I started working the other end with a stick and my hand. Suddenly Betty hollered, "Watch out!" The dam started moving with me on it, hunched over inside the culvert. I made a grab for the lip of the pipe with one hand, and for a piece of railroad iron imbedded at the mouth of the pipe with the other, as the dam washed from under my feet. There I was, hanging on as a five foot wall of water tried to push me through the culvert. Betty grabbed my collar and held on while I clawed my way onto dry land against the force of water. Without her help I couldn't have held on and would have had a white water trip through the pipe. I know I'd have been a battered and bruised hunk of humanity. And, you know what? March is too early to go swimming in cold mountain streams.

Betty is some understanding woman, wife and sweetheart. As I lay there soaking wet, with my heart pounding, we both had to laugh at the predicament I'd been in.

A trapper could rid the property of the beavers I'm sure, but no, I'll keep on tearing out their dams as they build them, hoping the next one will be built in a place acceptable to both them and me.

ANDY

Some of my most pleasant job memories are about Andy. He was a hard worker, and a real gentleman with a keen sense of humor. Most of his time was in timber management work, all over the Shasta Forest. In the fall I often worked with him, running lines, marking and cruising timber, marking experimental growth plots and occasionally collecting pine cones. He really seemed to enjoy his work, and took pride in doing it well. Things were not quite so complicated in the late 40's and early 50's. Environmental impacts had never been heard of, ecology was a word in the dictionary, but not in common usage and the committee approach to planning had not yet gotten off the ground. The results of Andy's management would be totally accepted today, but the methods and plans were all in his head. Believe it or not, it was a good system, with low overhead assessments. Andy and Vance, the forest timber management officer, were the only full-time employees in timber management on the Shasta.

One of my remembrances of Andy was near Lake Britton. We had just found a section corner down near the lakeshore. As we were checking our field notes, a commotion above us attracted our attention. A huge bald eagle was screaming at us and making passes about 10 feet over our heads. One of the bearing trees had a nest in the broken top. At least two young eagles were peering over the side. These were the first wild eagles I'd ever seen, and the thrill of it still lingers.

Andy's sense of humor never left him even when the joke was on him. He had just purchased a ball point pen, which was a new invention in the late 40's and one cost about $14. (And they weren't nearly as good as the present day 49¢ model.) He had shown me the pen and told me all about it the night before in our room at Fall River Mills. "It'll even write under water," he informed me proudly. On this particular day we were running lines in the flats near Burney, and had to cross Clark Creek. Andy was in the lead, pulling the tape, and decided to cross the creek on a log about eight feet above the water and over a deep pool. About half way across, the log snapped. Down went Andy, log and ball point pen. I rushed up to see if he had been hurt, which fortunately he had not, but he was standing in shoulder deep water. I suggested he try the ball point under the water, at which time he threw the pen, notebook and a few choice words at me. Now, Andy also had a speech impediment which caused him to stammer. I then told him he should always stand in water to talk, since his voice was perfect as he threw the objects at me. Even as we dried him out he could see the humor in this incident. Every time I worked with him for more than a day, I would find myself stuttering, which tickled him no end. He told me that my stuttering with a southern accent was far worse than his problem.

Andy and I worked on a fire team together for several seasons, and we really became good friends over the years. He visited me in Weaverville in the fall of 1958 and we had dinner and a real good visit. A few weeks later the most depressing job related incident possible happened. Andy, and a crew of firefighters were killed on a fire on the Cleveland National Forest. My remembrances of him will always be treasured.

LAW ENFORCEMENT BY THE GREEN FUZZ

Smokey Bears, Smokies, Green Fuzz, Tree Fuzz, Green Gestapos and myriads of less complimentary terms have been leveled at the Forest Service law enforcement arm. Just like other law enforcement agencies, we have done a necessary job; sometimes poorly, sometimes well; sometimes effectively and other times, not so effectively. Much flexibility has been allowed in Forest Service dealings with law violators. District policy varies, even on the Shasta-Trinity Forest, from issuance of citations to all people for all offenses, to a system of letting the enforcement officer decide what law enforcement measures are necessary. The latter was the policy used my me, while on the Pit, Weaverville and Mt. Shasta districts, though not without some criticism, even among my own employees. Their thoughts being that I was acting or allowing them to act as judge and jury, as well as enforcement officer. They were right, but I have always felt that the enforcement officer has a better "feel" for the circumstances involved, and his decision is based on better judgment than a disinterested judge, who just interprets the law without always understanding the facts of the case, or the thoughts and intentions of the offender. I agree that deliberate violations must be handled through the judicial system, but where ignorance of the law or regulation is involved, the violation can be handled administratively more effectively than through an arrest or citation. After all, when I can't remember how all the laws read, how can I expect the citizen who doesn't work with them to.

Over the years I've been involved in many interesting and varied cases. Let me tell you of a few.

My first fire law violation case was an unattended campfire which had actually burned through duff until it was outside the fire ring. It was in a large hunter campsite, with several vehicles and tents. After waiting an hour or two with a witness and no one showing up, I put the fire out and left a note for the owner of a certain vehicle parked at the camp to report to the Forest Service office in McCloud. He did, and I issued him a notice to appear in court for leaving a campfire unattended. The next day, a prominent citizen of one of the local towns appeared at the supervisor's office in Mt. Shasta, making a magnanimous statement that it was his camp where the citation had been issued, and he would assume the full responsibility for the supposedly unattended campfire. He further indicated that when he returned, the campfire was exactly as he had left it in the early morning, and if it was burning when I got there, someone else had been using their camp during the day. After a while, I felt more like the accused than the accuser, and that I should be ashamed to think that such a fine local citizen could be responsible for such a dastardly deed. We tore up the citation, and cancelled the date to appear in court. "After all, Bob, he has lots of influence in the county," I was told by the wheels in the supervisor's office. Nothing like this ever happened to me in the dozens of later cases throughout my career, but it did affect my willingness to issue citations for all the following years in law enforcement.

Mr. J. bought a few and stole lots of Christmas trees. His system was something like this. He would get written permission from small landowners to cut Christmas trees on their lands, which often as not, didn't even have Christmas trees on them. Then he'd cut wherever the trees were good and plentiful, usually on Forest Service lands, hoping not to get caught. No telling how many times he was successful, but eventually word got to the Forest Service in Weaverville that someone was cutting trees on Forest Service land in the Connor Creek country. We checked and sure enough, there were lots of fresh white fir stumps glistening among the remaining trees. By following a few clues picked up here and there, we ran down Mr. J., who had a bill of sale from a private landowner, plus a truck load of white fir trees. Everything appeared in order, except that I remembered the area owned by the private landowner to be a pure Douglas fir stand. It was night by now, but Frank, the acting ranger and I went to check the area from which the trees were supposed to have been cut. Sure enough, there was not a white fire on it. Not only that, there were not even any fresh tree stumps on it, so we confiscated the load of trees. The next day, with no trouble at all, we had matched dozens of trees on Mr. J.'s truck to stumps on the Forest Service lands in Conner Creek. We had Mr. J. dead to rights. He was kind of a likeable old rascal, and even after being caught red-handed, he complimented Frank and me on our efficiency. His tale of only doing it so his family could have a Christmas like everyone else was enough to make a grown man cry. It must have touched the judge, too, as his sentence after being convicted of a felony theft was light enough to encourage future tree theft attempts. Anyway, Mr. J. was not very ashamed of his act, because about April of the following spring, I'd stopped at a little cafe down river from Junction City to buy lunch, and there was Mr. J. at the counter with several other people. "Hi, Bob, have you caught any Christmas tree thieves lately?" he asked in a loud greeting. I sat down next to him and ate, as old friends should do.

Let me tell you how trees are matched to their stumps for identification. The growth rings and characteristics or blemishes on the stump will match perfectly with those on the tree butt, especially if cut with a handsaw, and usually there is a torn tag end of bark left on the tree, which fits perfectly into the place it is torn from on the stump as the tree falls. This perfect match is as acceptable as fingerprints in court in identifying stolen trees.

Another Christmas tree theft was discovered on the north slope of Mt. Shasta because the cutter accidentally strayed over a property line onto Forest Service land. I'm sure the accidental straying was just that, as the cutters were working right along the road, and were culturing, pruning and thinning as they cut trees. They would hardly go to this extent if the trees were being swiped. They assured me they had not read a K-tag properly in determining the cutting lines. Nevertheless, this accidental trespass led me to nose around the area long enough to find lots of scattered, high-graded stumps on nearby national forest lands. This led to a tedious search through hundreds of legitimately cut trees before finding about twenty trees to match the scattered stumps on Forest Service land. Again, the conviction and sentence were hardly enough to dissuade the cutter from continuing his thieving ways. Perhaps his loss of contract with the Kimberly-Clark Corporation did more toward making him an honest man than the light fine did.

Oh yes, he asked me if I'd be interested in working for him the next year, just to keep him out of trouble.

During the mid 70's, strict state and county laws, tougher enforcement and use of certified permits has made it more difficult for people to get away with Christmas tree pilfering, but it probably won't ever stop it completely.

Fire camps are notorious for coming up short of supplies after the fire is over and the camp is broken up. Lots of stuff is stolen, much is carelessly broken or destroyed, some is worn out, a little bit lost or burned on the fire and a lot of it goes to Forest Service warehouses other than where it originated. Forest Service employees are sometimes guilty of having "taking ways" on fires, and some of these have been apprehended with the result of the employee losing his job and even receiving a jail sentence.

I'm sure the conditions are no better or no worse than they were thirty-five years ago, no matter how hard the Forest Service tries to remedy the problem.

I believe it was in 1946 that we had a fire just outside of McCloud. A fire camp was set up with all kinds of good things to eat and many of the comforts of home, like sleeping bags and cots.

As camp was breaking up, Paul was looking around the area behind the campsite to make sure everything was clean and orderly. He discovered a carefully camouflaged pile of groceries and sleeping bags stashed under a big log, obviously there to be picked up later. We set up a stake-out after the camp was gone, and late that night the fire camp cook showed up to gather the loot he'd carefully hidden away.

Just after the mid 60's, and partly as a spin-off from the Haight-Ashbury District of San Francisco, a new sub-culture section of society began to descend (or ascend) upon the Shasta-Trinity National Forest, especially on the Sacramento Ranger District. Mt. Shasta has always has a magnetic effect on many persons and groups for different reasons of religion, mysticism, superstition, legend and lore. Its majestic beauty affects most everyone, including me.

These modern-day nomads began to trickle into the area in singles, pairs, and small bands, with shocking effects on the local citizens and especially the local police and Forest Service. They came into town, dirty, barefoot, bearded, often smelling, wandering the streets and entering stores and restaurants, much to the dismay and displeasure of the merchants. They were accused of stealing and contaminating everything they touched. Signs appeared in store windows—"No Hippies Allowed." Indeed, many were highly undesirable types, but at the same time, many were well educated, intelligent, and even had money to spend. Afoot, in beat-up old cars and vans, hitch-hiking, and occasionally with a pack donkey, they would arrive. The effects of drug use and abuse was evident in many, as they showed up at the Forest Service front desk in Mt. Shasta, looking for maps and information about "The Mountain" and the whereabouts of their friends and fellow travelers. On occasion they would be glassy-eyed and incoherent from dope and malnutrition as they talked to the receptionist, who oftentimes had to call for assistance from me or others in the office to interpret their questions or to decide whether or not to ask the local police to pick them up. Some were very pathetic, mixed up people, groping for something that was not there. For several years, even 'til now, these people have continued to come and go, at times causing problems with and for the Forest Service.

Harold, the police chief called one morning, "Bob, a young newly married couple stopped by to report a nude male hippy wandering around the Panther Meadow campground, do you want to take care of it?" Off to the mountain I headed. It was about 10 o'clock, and still pretty cold at the 7,000 foot elevation, but just as the people said, there was a bearded, long, lean and naked young man nonchalantly walking from his camp toward the water hole, with the neighboring campers paying little attention to him. "Oh yes, he's been doing this for a couple of days," they said.

"Put some clothes on," I told him, "we're going downtown to see the judge."

I had to take him to jail in Dunsmuir, where he was booked for indecent exposure and held for a couple of days awaiting a hearing by Judge S. Meanwhile, I got a phone call from the district attorney's office telling me I couldn't arrest a man for running around nude in a campground, or anywhere else, for that matter. "You mean," I said, "that if a man were walking down Main Street in Mt. Shasta with no clothes on he could not be arrested?" "That is correct, Mr. Gray," answered Mrs. S., the D.A. I was speechless, as she went on to say that a nudie would have to be doing something lewd or obscene before he or she could be arrested in that the nudity had to be offensive to someone, and that mere nudity was not classified under either category. So, she advised me to drop the charges.

I went to the judge, telling him what I'd heard from the D.A. He was a little upset, but told me the charges shouldn't, and would not be dropped. Well, we went on with the trial, but the D.A., who was supposed to be the prosecuting attorney, didn't show up. This didn't faze Judge S. one bit, he went right along with the hearing, with the public defender representing the defendant and the judge acting both as prosecuter, judge and jury. Needless to say, we won, but I never really felt good about the verdict under the unusual circumstances at the trial. During the trial, I was afraid the public defender was going to ask me if the defendant was the man I'd seen parading through the campground in the altogether. I think I'd have had to tell him, "No, sir, I don't recognize him with his clothes on."

Up at Cliff Lake one day I found no less than five unattended campfires, each of which was surrounded by litter, garbage, and pretty scruffy camping gear. Obviously the people intended to come back, so Jack and I decided to wait for the campers to return. Within an hour, five very unkempt young men and a near equal number of scantily clad young "ladies" appeared at the camp. There were no campfire permits, and none of the group were particularly willing to admit to owning the fires, though they admitted to owning the junk scattered around. While writing citations to the men, the ladies decided to change the few clothes they had on. Well, I busily kept writing, while Jack kept his eye on the girls. A very disconcerting situation to cope with in a businesslike manner, but Jack and I prevailed.

Sometimes troubles come in bunches. It was the 4th of July evening when I overheard on my Forest Service radio that Larry was having some problems at Lake Siskiyou with a group of campers. There was a religious festival, with a circus size tent and hundreds of people at the campground. This large, well organized group had been, according to them, harassed by a lesser size, but highly zealous group who called themselves the Christ Family. The philosophy of the two groups was quite different. The large group was obviously made up of clean, intelligent and probably well-to-do families. I judged this by their clothes, camping gear, fancy motor homes and high degree of organization. At the same time, the "Christ" family were barefoot, smelly, and claimed to own nothing personally, but with God, owned everything.

The trouble for the Forest Service started when Al, the superintendent of the campground reported an unsafe campfire on private land, with a bunch of odd people sitting around it, who refused to leave when he asked them to. By the time I arrived at about 10 p.m., there were Forest Service vehicles, California Highway Patrol vehicles, Siskiyou County Sheriff's vehicles and Mt. Shasta City Police cars, all with red lights flashing, and nobody doing anything. It looked like a riot was about to take place. The campfire was glowing brightly about 100 feet from the road, and twelve hooded, robed figures were huddled around it, paying no attention to the commotion they'd caused. No one in all the law enforcement agencies present had talked to them, so I walked over to the campfire, followed by all the other officers and spectators.

"Who is the spokesman for your group?" I asked no one in particular. That's who answered—no one. I tapped a huddled figure on the shoulder. "This is private land and you do not have permission to be here or to have this campfire. What is your name, please?" Still no answer, except a female voice across the fire from me said, "This is God's land, it does not belong to anyone else." "Are you in charge here?" I asked her. No answer.

"Look," I told one of the men, with my nose close to his, "I need a name to put on this violation notice for building a campfire without permission. What is your name?" After a pause, "We are the Christ Family, and don't have names," he mumbled.

My patience was wearing thin, "If I don't get some identification and a signature on this violation notice, I'm taking you all in to jail." "We're the Christ Family," is the only answer I got.

We put out the campfire, and the twelve men and women obediently walked over to the pickup where we loaded six into the back of Larry's rig and five into the back of mine. Yes, I lost one of them somewhere along the way. It was a long way to the county jail in Yreka in the back of the pickups, but honestly, their super B.O. was enough to make your eyes water, so I'm sure the open air trip didn't hurt any of them.

At Yreka, before unloading at the jail, I gave them one more chance to identify themselves and sign the violation notice so they wouldn't have to go to jail for a misdemeanor violation. They still refused to comply, so we marched them into the jail where they were given clean clothes and told to shower, which they did without protest.

By the time I had signed all the papers as arresting officer, one of the prisoners, a young woman, started talking through the bars about the evils of government employees and law enforcement officers in particular, and expounding on the virtues of Christian principles. Among these, she listed no violence, no material possessions, no sex and no discrimination of race, creed or religion. Even in prison garb, with her freshly washed hair and face, she was a very pretty young woman who, I think, sincerely believed in the crusade in which she and her fellow clan were involved.

The sheriff's department had better luck than I in identifying the "Christ" family. During the court appearance three or four days later, they were called by their earthly names by the judge, as he heard their plea of Not Guilty. They accepted a non-jury trial, and were sentenced to four days in jail, which had already been served. After the trial, several of the poor misguided souls (my opinion only) thanked me for my sympathetic understanding in handling their case.

Don't lose your cool! That's very important in law enforcement, and on one occasion I did just that. It was near Wagon Camp, a favorite camping area for the so-called hippie culture. There were several camping spots around the meadow and along the small stream. Each was within sight of the next and on this particular day, about a dozen people were sitting around a small campfire at the first site I approached. We made small talk in a friendly fashion, about fire prevention, sanitation and the weather, before I walked through the woods to the next several campsites. There were two smoldering fires, both unattended, among the sites. I knew the owners had to be in the group of people I'd seen at the first site, so I went back and asked, "Who has the third and fourth campsites down stream from here?" I foolishly told them about the unattended fires of which, I'm sure, they were already aware. "No, we don't know who was camped there, except that they left yesterday." Now, I hate to be lied to, and I let them know it, as I lit into them about their unsanitary conditions and their disregard for fire safety. They listened quietly, as I raised my voice in frustration and anger. All the time I talked, one of the girls kept scratching her stomach, causing her ragged blouse to move about, creating a peek-a-boo effect to portions of her upper torso, exposed as the holey materials moved back and forth in time with the scratching.

Finally I left, and when I got to my pickup, it wouldn't start. My battery was too weak, and I couldn't even use the radio to call for help. You guessed it. The whole bunch came over and pushed me to get started, grinning and full of love for their fellow man. Quite embarassing, after the chewing out I'd given them.

A week later a friend of mine, working for the F.B.I. said to me, "Either you didn't recognize me, or kept it well concealed when you were talking to the hippy group at Wagon Camp last week." Seems he was working on a case, and was disguised as a member of the group I was talking to. The fact was, I did not recognize him.

Stealing and shooting Forest Service signs is an avocation of many people—young, old, rich or poor—it makes no difference. A rifleman just had to dot the "i" or put a hole in the middle of the "o" in a new sign, while the scatter gunner has to test the pattern of his shotgun on the large paper poster signs and metal signs along the roads. If they don't shoot 'em, they steal 'em—selectively, though. Who could resist taking a neat, wooden, routed sign that says *Nit Wit Camp,* or *Nine Buck Butte.* Even the *No Camping* and *Garbage* signs are popular for the walls of kids' rooms. Not too much time or effort is spent looking for the small signs, but a few years ago, a large expensive, metal map of the Shasta-Trinity Forest disappeared from an informational type sign at the Everitt Vista on the road to the Mt. Shasta Ski Bowl. It was worth several hundred dollars, and had just been placed on the site a few days earlier. I checked the area to see if it might have been tossed in the brush nearby, but could find no sign of it. Since the value of the theft made it a felony, the district ranger wanted to call in the F.B.I. to help solve the crime. I asked him to hold off for 24 hours, because in my search around the area I'd discovered freshly carved initials and names in the wooden benches beside the sign. The names and initials matched those of kids I knew in the local high school, who happened to be friends of my son and daughter, also in high school. No one in their right mind would steal something and then carve their name in the bench beside it, would they? I thought to myself. Yes, they would, I concluded, if they were thoughtless, typical high schoolers. After all, I should know—I had two of them myself. First, I checked carefully to see if the initials D.G. or M.G. (my son and daughter) were among those carved on the bench. They weren't, much to my relief. I went directly home and told Doug and Mary, "Someone stole a $500 metal sign at Everitt Vista today, and the initials of K--, D--, B--, J-- and a few others were carved in the bench nearby. Do either of you know anything about that sign?" They looked at each other and then blankly at me. "If that sign is not back in place within 24 hours, the F.B.I. is taking over and there's going to be trouble with a capital T for some Mt. Shasta high schoolers."

In about five minutes Doug left, mumbling something like, "I've gotta get some binder paper," with Mary close behind.

I really don't know who had it, but in about two hours, my phone rang, and a voice said, "Mr. Gray, the sign is back in place," and then silence as the caller hung up. Doug and Mary never commented on its prompt return.

CHRISTMAS TREEING

Among the fringe benefits of being a ranger until the early 70's, was the privilege of getting a free Christmas tree for my family, an elderly friend, a sick friend, or a needy family. Paul, the forest supervisor usually needed about a hundred or so for hospitals, schools, organizations and the like in the Redding area, and the job usually fell to the Sacramento District to provide these trees, since we had good access and probably some of the prettiest silver tips in the state. We would organize a four or five man crew, chain up the stakeside, and armed with handsaws or small chainsaws, head for the Christmas tree area selected for the cutting. In addition to

Paul's hundred, another seventy-five were cut for district personnel, local schools, hospitals and churches. It was always a fun day, usually around Thanksgiving or earlier when we went out. If snow was already deep in the area, we took the snow cat to skid the trees to the trucks. Everyone put a ribbon or special mark on his personal tree to keep it from getting mixed up with all the others. Silvertips grow above 6,000 feet, and are the most symmetrical, evenly limbed and handsomest of all Christmas trees. Their excessive weight and stiff branches make them difficult to load very tightly, but are prized by everyone.

Nowadays, even the Forest Service personnel must pay a one or two dollar fee for a Christmas tree permit, as does the public, who used to get free use permits for a family tree.

Increased local population, commercial theft and more demand for trees has created a need for closer and tighter controls on all timber functions, from the administration of timber sales to issuance of Christmas tree permits. Consequently, regular patrols are made by fire prevention technicians to insure compliance of the rules of cutting Christmas trees. To me, the little bit of revenue obtained from tree permits does not nearly repay for the loss of goodwill and free family fun for families who feel the forests are theirs, and should, therefore, be entitled to a tree without cost.

TRAINING

The first half of my career was self-training; training by doing; training by making a mistake only once; and training by watching someone else doing something without letting him know you were watching. Then, the second half saw the training pendulum swing to the other extreme, with formal training sessions sometimes taking too much time for what they were worth.

Don't get me wrong, much of the training was good to excellent, with high caliber instructors and materials. Some could have as well been left undone. In fire control, we were trained and are still being trained in checking spark arresters on equipment. Good—you bet, but not two or three days of it. And the same people go, year after year.

How about Advanced Fire Behavior? Another good topic, but even here, should it take eighty hours to learn that dry sticks burn faster than wet ones?—or that fires goes uphill faster than downhill? or even that grass burns faster than logs or trees?

Well, anyway, training has been interesting, dull, useful and otherwise; and sometimes "training for training's sake."

Training for trainers, or Job Instructor Training was mostly good in helping develop leaders and instructors. Emphasis on visual aids was stressed highly—indicating that what you "see" registers more than what you hear or read. Irwin, from the Six Rivers gave a five minute instruction session on how a smoke-jumper gets out of his jumpsuit and gear after landing on the ground. He had a "smokejumper" fully dressed in padded suit, helmet, gloves, face mask, and the works. At a given signal, the jumper started unzipping, unbuttoning and taking off the bulky gear. At the last moment the whole outfit was flung aside, and there, standing in front of the thirty man class was a very feminine, very blonde, very sexy looking damsel in the flimsiest of female apparrel, thus proving the effects of visual aid materials. I'm sure none in the class have forgotten this bit of instruction, as I'm equally sure no one in the class remembers anything else taught that day, proving that visual aids can also be overdone.

It was in 1952 that I became driver-trainer for the Shasta Forest, and it happened that Vance, already a legendary forester on the Shasta, was my first trainee. We pulled out from the front of the supervisor's office, drove a hundred feet and turned right at a stop sign—without stopping. I said nothing, but made a note on the form. Two more stop signs without stopping, and I said, "Vance, don't you see those stop signs? You've run right past three of them." "Yes," he answered.

"Well, why don't you stop??" was my puzzled reply. "Now listen, young fella, I've been driving for thirty or forty years, and I haven't ever stopped at those signs unless I had to. I'm not going to start it now just because you're here."

I erased a few check marks. After all, Vance did have to drive in his job, and he really did look both ways before running the signs. Yes, I was a little more firm with other trainees. Vance promised not to have any wrecks, and he didn't.

"DEAR DIARY"

We might not have known what we were going to do tomorrow, but we sure knew what we did yesterday, last week, last month and last year. From the time I started work until the mid 60's an Official Diary was kept by all permanent Forest Service personnel. Most everyone hated it, but I sometimes found it beneficial as a reference to completion dates of jobs, dates of important incidents, and especially in writing this book. A few excerpts give a sampling of a "typical Bob Gray work day."

3-5-56 Took Stouts Mdw. snow survey. George Manes and I skied to meadow in blizzard. Got lost and pooped out. Reached cabin at dark! Had troubles crossing Shady Gl. & Tate Cr. 12 hours.

3-6-56 Took survey in a.m. Maximum depth 164". Average 150". Broke sampling tube and didn't get away from meadow until 1230. Arrived at pickup at 1900. Battery dead—so we walked to Nursery. To McCloud in Lanquist's pickup. A tough trip. 13 hours.

3-13-56 Conservation movies and talks all day. Presented at McArthur High School, Pondosa Elem. Sch., Pondosa PTA & Fall River Lions Club.—Got sort of hoarse.

5-25-56 Trimbles copter took Del and me and Waller plus two loads of gas and generator and supplies to Weaver Bally. Spent most of day getting lookout oriented and operating.

Homer picked me up at 1500. Helped Waller work on office radio 'til 1730.

7-10-56 Packed trail crew to Morris Mdw.—Thunderstorm in p.m. Rain for 3 hours. Got soaked to skin. Helped set up camp and lined up job for crew for next 10 days.

8-17-56 Office at 0800; trail plans, payrolls, weather, etc.

1500—Took stock truck to Bowerman Mdw. trail to pick up Bernie, Bob, and trail contractor. Met Homer.

1650—Granite Peak reported smoke at Keno. Homer and I left for fire in his pickup. Arrived at fire about 1800. 5 acres when we arrived. Minersville, Weaverville & two state crews at fire. Scouted it in copter. Completed dozer line at 2200. Remained on fire as fire boss.

11-20 & 11-21-56 Went to Deep Creek and helped crew on bridge job both days. Safety talk included. Peeling and skidding logs. Slow going, but if weather holds, we'll finish the darn thing.

12-1-56 Attended funeral services of Anderson and Maxwell. Note: Had I not been camped at Deep Creek, I'd have been with Andy and Forrest on the Inaja fire, as I was called by Pettigrew to go to the fire.

12-31-56 Took annual leave and killed a bear with the McKnights.

So ended the 1956 year, A.D.

MORE ON HOW FOREST FIRES START

"The Friendly Southern Pacific"—It has probably done more toward keeping the Sacramento (Mt. Shasta) Ranger District crews sharp, efficient and on their toes, than any training program ever dreamed of. For several decades, it has been the biggest single cause of fires up and down the middle of the Shasta Forest. Fire crews love those railroad fires. They create excitement, sometimes come in bunches, break the monotony of waiting long periods between other fires and really give lots of on-the-job training in use of pumps, hoselays and handtools, plus bringing an occasional air drop. Even the burning lumber cars and boxcars loaded with who-knows-what, add a new dimension to the run-of-the-mill fire. Speculation runs rampant over the contents of those burning railroad cars. "Does it have toxic chemicals?—Does it have explosives?—Does it have radioactive material?" No, it turns out it's just a load of farm machinery, or cornflakes. But, on occasion, it is a hazardous load of flammable or toxic material, and the crews really earn that "hazard pay."

A burning flatcar can spread fire for miles along the tracks, with as many as 25 to 30 individual smokes in ties, chips, grass or whatever is in the right-of-way. One or more might spot into the trees or brush outside the cleared right-of-way, and develop into something more than a nuisance type fire.

During the early 60's, woodchip shipments became a major business with the S.P. Company and all major West Coast lumber companies. Hundreds of open top cars, heaped and overflowing with wood chips were shipped through our area. It looked like a miniature blizzard of flying chips as these cars sped down the tracks enroute to processing plants. Within three years, the chips had piled up several inches deep alongside the tracks, and the incidence of "chip fires" ballooned to nearly a hundred on our district alone. It got to the point that we could expect at least one fire every afternoon in the five to six miles of track between Upton and the Black Butte overcrossing. All westbound trains began braking to reduce speeds before reaching the controlled speed area near Mt. Shasta, and the hot metal flakes off the brakeshoes would easily ignite the tinder dry chips. The fire incidence got so high in 1964, that we parked one of our tankers at the Deetz crossing every afternoon from about two to four p.m. so we'd already be there when the inevitable fire was discovered. From the time a hot brakeshoe flake hit the chips until smoke began to show was less than one minute. The smoke would sometimes show before the train got past, and the wind from the train would immediately fan the smoking chips into flame. In June of 1965, the S.P. Company, under my direction, burned the accumulated chips on both sides of the tracks for a distance of two miles each side of Deetz Crossing. About the same time, all shippers were required to put a heavy netting over all chip cars to prevent the chips from blowing and shaking off. Immediately the fire incidence was reduced to just an occasional fire and has remained low for the years since then.

Working with the S.P. Company was a good experience for me since they were, and still are a cause of concern for wildland fires. Much of my time was spent in working with S.P. officials. I developed a close working relationship with them and felt they cooperated well with the Forest Service, on our district especially. Each spring a meeting and fire inspection trip was planned and made to cover the right-of-way from Redding to the Oregon line, through areas protected from fire by the State of California and U.S. Forest Service. The inspections in the early 60's were "an experience to remember." The company still maintained and operated the "Shasta" car, a holdover from the plush passenger era of railroading. It was not a chrome, flashy modern passenger car, but an elegant, dignified, solidly constructed old timer. It was used by division superintendents, visiting dignitaries and the upper eschelon of railroad personalities, with a dining compartment, a kitchen with a white clad black man cook; chef's hat and all. The club room had deep velvet upholstered seats and mahogany trim, but best of all, it had an observation platform on the back, like Al Smith must have used in his presidential campaign in 1928. With cigar smoking S.P. officials, we made fire hazard inspections in the Shasta car in 1960, 1961, and 1962, but amid all this luxury, cigar smoke and delicious eating, not too much was accomplished in the way of inspecting. We usually had to make additional spot checks of certain areas to see what we had missed from the Shasta Car.

Well, this era ended when the S.P. mothballed the Shasta, and the next year we made the inspection from a speeder. It was some letdown, but the inspection was much more efficient, as we could stop and look at things and were put on the siding a few times to let the freight trains by, enabling us to discuss what we had seen. The level of the railroad officials within their organization dropped a little when the change to "speeder inspections" took place, but the people like Al, Gary, Bob E., Paul, Nick and others were people we could deal with directly in getting compliance with our requests for hazard reduction work. I found that reasonable requests got reasonable results from the company, and that outlandish requests got nothing. Railroad fire occurence has dropped from over 100 fires in 1964 to about 10 per year in the past years. It is a record that can be improved still, and the railroad is working with the Forest Service toward a mutually accepted goal.

The past few years, a "high-rail car" had been used for making railroad inspections. It is a standard sedan or station wagon equipped front and rear with flanged iron wheels to hold the vehicle on the track while the rubber tired wheels run smoothly on the rails. This is the current, and the best method of making inspections. Quietness, speed and comfort, as well as good visibility make the high-rail car an ideal way to go.

A few years ago we were waiting clearance at Black Butte siding before proceeding with the inspection. Bob E., the new assistant division engineer for the company was among the inspection party. Since we had about twenty minutes to wait, he asked me to show him how to identify some of the trees in the area. "Sure, come on, we'll go over to that clump of trees and I'll give you a lesson in tree identification." I could see that there was a mixed group of conifers, so we walked over to the trees. We had the lesson, and were ready to go back to the high-rail car when I took an exciting jump off a rock.

I was standing on top of a four foot lava rock, making my way back toward the tracks when I jumped into a small clearing between two clumps of brush. Just as I jumped, a rattlesnake slithered into the same opening right where I was about to land. I'm sure I hung in the air for a second or two. At least I tried to, but I still landed astraddle that snake. Naturally, I lit running. But that wasn't the end of my tale. About three steps further, my front foot landed on an old barrel hoop, flipping it up so that my other foot got caught in the hoop. I took a header into the brush, but hurt only my dignity. Now all this time, Bob and the other three men were watching me, but didn't know about the snake, so couldn't understand what caused all my gyrations. By the time I recovered my composure, the snake was gone, and when I told my story to the men waiting for Bob and me at the high-rail car, they all agreed that for some reason or other, I had hung in the air for a while after jumping from the rock, and the tumble after hooking the barrel hoop with my toe was of classical form and grace.

And this story is still not over. We left Black Butte, and in a few minutes I discovered my glasses were missing. They were lying just beyond the barrel hoop!

Chain saws start fires—Exhaust sparks from tractors start fires—friction from cables start fires—even a spark off the track of a Cat has been credited with starting fires. But I know that careless smoking causes most of the fires in logging operations. Despite warnings by bosses, T.V. commercials, no-smoking posters, and harassment by the forest rangers, loggers continue to smoke. I've caught several of them, and have in all cases taken criminal action against these offenders, because *they know better!* Once while driving through a Forest Service sale area, I could see a faller working about a hundred feet from the road. He was falling a large tree and hadn't seen or heard me because of the noise from the saw. I could see a cigarette in his mouth, but couldn't be sure the smoke I saw was from it or the chain saw. I started walking toward him and hadn't gone far when he saw me coming.

His actions were pretty cool. He kept right on sawing, but turned his head a little bit and cropped the cigarette from his lips, covering it with sawdust with his foot as the sawing continued. I still wasn't sure it was a lighted cigarette I'd seen, but his action made me more suspicious. I walked up to him and he shut the saw down. We talked a few minutes, before I started digging around in the fresh sawdust with my foot. He stopped talking a moment, then said, "It should be about here," as he dug out the still smoking butt. "Do you always smoke while working in the woods?" "Yes, but I've never been caught before. You must have good eyes to have seen that cigarette from the road."

He appeared before the judge in a few days where he got a fine of about $30, but his worst penalty was that the logging company canned him. My problem was that I don't think the fine or the firing did anything to keep him from continuing his bad smoking habits wherever he may be working.

Lower Slate Creek is inhabited by a group of people who have chosen to live in a way far removed from the modern ways of society, seemingly content with fewer luxuries, and a standard of living comparable to the early 1900's. This is not an adversely critical appraisal of these folks, as I think they're hanging on to a way of life that most of the rest of us might like if it weren't for the Jones who we feel compelled to keep up with. They seem happy and content.

Mr. G. is one of the older inhabitants of the area, probably in his eighties, with whom I have had many interesting conversations. One day as I passed the old abandoned Slate Creek School, he was sitting on a rock smoking his pipe in the weed covered playground. About an hour later Rosie on Sugarloaf reported a fire by the old Slate Creek School. It had obviously started by the rock Mr. G. had been sitting on. I went up the road to his house after the fire was controlled to tell him he'd started a fire where I'd seen him sitting a couple of hours before. "Now just a minute, young fella, I didn't start no fire, and you shouldn't be spouting off about something you don't know," he told me in a slightly ruffled tone. "Well, you did, too, and you know you did, and I oughta' write you a citation," I told him. I really didn't want to, but I did want him to admit his guilt, because I figured I could handle the case so he wouldn't have to pay suppression costs, and still close the incident as a solved trespass.

A few days went by, and I kept hearing rumors that some kids had started the fire, and after a little bit of investigation I found out that was the case. All I could do was go apologize to Mr. G., which I did very.meekly. "I'd have told you who started it if you hadn't been so ornery when you were here," he told me as I left.

Arsonists are a plague to firefighters, and fortunately for me, there have been relatively few incendiary, or arson fires while I've been fire control officer on any district. Incendiary fires are started by many methods, some simple, some crude, some complex, some unique. Among the unique methods, one was discovered in the Pondosa area during the early 60's. It took a little thought, planning and ingenuity to come up with this idea, which used concave vanity mirrors to concentrate reflected sun rays on flammable fuels. For several weeks fires had been reported along Highway 89, usually no more than 100 feet from the road. In each case, the scorched remains of a mirror was found at the origin. Sometimes the heat had melted, twisted and broken the mirrors until they were hardly identifiable. The method was to place the concave side of the mirror facing southerly, with the focused rays pointed into a punky log or dry litter on the ground. The focal point was about 24 inches from the mirror. Just as with a magnifying glass, the concentrated rays would ignite the fuel, which would smolder in the punky logs for undetermined periods of time before bursting into flames as weather and fuel conditions dictated. It was never determined whether the mirrors were "aimed" at a certain spot or were just placed approximately, hoping that sooner or later the sun rays would deflect onto the right spot as the earth rotated, or the earth's position was tilted to the right angle due to seasonal changes. After the 4th or 5th fire, the Forest Service searched the areas along Highway 89 in the pattern the mirrors had been placed. Two or three more "sets" were found during the search, but at least one was missed, for during the following summer, another mirror fire was discovered. The mirror had failed the first time around, but the next year it did its duty, even after a winter of deep snow. An interesting pattern showed on one of the sets found before it started a fire. A scorched mark was inscribed on a log where the focal point had inched its way across each day as the sun moved across the sky. Who knows, there may still be a mirror out there aimed at an old stump, ready to ignite when the time is ripe.

Who thought up this scheme to start fires? We never solved the riddle. It could have been a disgruntled logger trying to get even with the company or the Forest Service, or even a Forest Service employee trying to create a little action for his crew, to liven up an otherwise dull summer.

Some fire investigations are simpler than others. I was sent to Red Bluff to talk to a man thought to have started the Sulfur Fire in Cottonwood Creek. He had been brought out of the remote canyon by Forest Service helicopter after discovery of the fire, and was taken to the hospital for minor injuries and shock after being lost several days. No one had yet talked to him about the fire until I approached him to ask questions. Before I could open my mouth, he started talking, "If you're going to ask how that fire started, I'll tell you right now that I did it, and am glad of it." He went into great detail how he had become separated from his hunting partners, and wound up falling over some bluffs into Cottonwood Creek, where he became totally lost. "For three days and nights I wandered up and down that canyon, not remembering where I had come into it. I built signal fires every day, and no one saw them. Even the airplanes flew right on by, so when I began to get desperate and down to my last match, I figured I'd build a signal fire that nobody could miss. First I picked the biggest brush patch I could find, figuring it wouldn't be any loss to burn, and set it on fire at the bottom. Sure enough, it wasn't any time until helicopters, smokejumpers and air tankers were right there. The helicopter brought me out, and here I am."

The Forest Service decided he'd done what any prudent man would do, and dropped the case then and there.

CAT TALES

During and after thunderstorms, lookouts quite often see smoke from fires that later go out from heavy rain or lack of fuel. These smokes may puff up intermittently, or show for a few minutes and disappear, or show up for several hours before disappearing for good. Occasionally, one of these smokes will dissipate, only to show up days or even weeks later with disastrous results. For this reason, it is important to thoroughly search out by aerial recon, or even on foot these possible "sleeper" or "holdover" fires.

Verna, on Slate Mt. reported a smoke near her lookout one day thusly, "It came up just over the brow of the hill from me, and a distinct column showed for several minutes before it disappeared."

Since all the other fires from this storm had been found and taken care of, I decided to take Bob F. to Slate and see if we could find the fire. My plan was to walk out to Budweiser Gap, about a mile and a quarter away, get on the back azimuth reading from Slate Mt. which was visible from Budweiser Gap, and walk straight to the lookout on the designated bearing. This way, I was bound to come upon the burned area, or even the still smoldering fire, though the base of the smoke was hidden from the lookout by the brow of the hill.

We had one problem which I'd failed to see, and that was Slate Mt. had no Shasta net radio, and I had no Trinity net radio, so the only solution for communications was for Bob to stay in my pickup at the foot of the 30 foot tower to talk to me, and holler up to Verna to keep me on the proper azimuth as I used a signal mirror to show her my location. With this plan, I left the lookout on foot, with a shovel and portable radio to search for the fire by myself. About 10 minutes and a quarter of a mile later, as I was struggling through a dense brush patch, a mountain lion sprang from the brush less than 5 feet from my side. Being startled was putting it mildly, as the lion bounded away from me, and when she stopped about 50 feet away and stood facing me, it was a real surprise. I really wasn't afraid, but found it quite interesting to be looking at a wild mountain lion.

We watched each other for several moments, it seemed, and soon I playfully called "Kitty, Kitty," to which the feline creature took a couple of steps toward me.

Since I wasn't really trying to entice her any closer, I then took a couple of steps toward it. It backed off a few feet, but still didn't leave. By then, I decided I'd better get on with my business, so I detoured around the animal, and was much surprised to see it following me at about 30 or 50 feet distance. Still I wasn't unduly alarmed, as there was no sign of hostility; just curiosity on the cat's part, though my own curiosity had long gone.

For about a couple of hundred yards that tawny creature followed me, with me continually looking over my shoulder. Just keeping track, you know. Suddenly, about 20 feet in front of me there was that cat again. No, because a quick glance behind assured me I was still being followed. There was a second, and much larger lion just ahead of me, and it looked far less friendly than the first one. Maybe my imagination was working, but this new critter looked twice as big and definitely had a scowl on his face, though he showed no real signs of hostility.

By this time, I figured this nonsense had to stop. I raised my shovel like a rifle and went "BANG, BANG!" They didn't budge, except that the second one circled behind me to where the first one was. At least I can see them both at once, I was thinking. They continued to follow, sometimes within 20 feet, stopping when I stopped, and walking when I walked. I sweated, fretted and prayed a little, before deciding to call Bob on the radio to tell him a couple of mountain lions were on my tail, and that I was going to keep walking to the Budweiser Gap clearing. Well, when you talk on the Forest Service radio, everyone hears you, especially when you don't want them to. In seconds everyone was aware of my plight, and the dispatchers were ready to send a copter for me. Lenore, on Sims, reminded me that once I had told her, "Don't worry, Lenore, he's as afraid of you as you are of him," when she encountered a mountain lion while on an evening walk.

I declined the copter, and continued to the Gap about a

half-mile away. My friends stayed with me all the way, but at some greater distances. As I reached the opening in Budweiser Gap, I was startled to see a man enter the clearing from the Incline Trail. It was Mr. Mardahl, a timber cruiser for International Paper. "Man, I'm glad to see you—What do you see behind me?"—"My gosh, a mountain lion," he answered, then he saw the second one as I asked him to look again. While we talked, the lions melted into the brush, but I was sure thankful to have a reliable witness to this unlikely tale.

I never did look for the fire anymore, and Mr. Mardahl decided he'd seen enough timber for the day, so we walked down to the road to his car and to where Bob picked me up. Yes, we looked over our shoulders occasionally on the way down the trail.

Back at the office there was a neatly wrapped and ribboned box of catnip on my desk. From my friend Bob M. and the gang.

SNO-CATTIN' AROUND

The thrills of Disneyland rides and the beauty of travelogue brochures are some of the pleasures of snow tractor travel. Of all memorable experiences in the Forest Service, the work with snow tractors has been foremost. Breathtaking beauty and hair-raising mishaps come to mind.

Tom always had the knack of showing up during the worst weather possible to fix the radio at the top of the chair-lift, or on Gray Butte. Sometimes both on the same day. One such day, with swirling snow, loose powder, and everything else needed for a disastrous trip, we headed for Gray Butte. About a quarter of a mile from the top, the narrow road is gouged out of the steep hillside and the deep snow pack had completely obliterated the cut forming the road. They only way to tell the road was there was by the absence of trees in a straight line ahead of us for about 400 feet. We stopped at a fairly flat place to take a look before deciding whether to go on foot or try to take the Thiokol the last quarter mile. The load would have been heavy to carry, and the warm tractor was pretty comfortable to want to leave. I figured a good operator like me could make it across by nosing the front up the hill, and "pouring the coal" to the machine, thus staying on the road while sidewinding across the slope. I figured wrong.

Hardly had we started when the snow cat started slipping downhill. I over-reacted and pulled the nose straight up the hill, at the same time, killing the engine. For a moment we hung on the hillside, then started sliding backwards down the hill. It was steep, like a roller coaster. I was pulling the steering levers and standing on the brakes. We kept sliding. Faster and faster we went. Tom was hanging on for dear life all hunched over waiting for something to happen. I didn't have enough hands and feet, and couldn't see where we were heading except downhill. Boy, oh boy, I hope we don't hit a tree, I was thinking, and I don't know what Tom was thinking. We kept sliding—straight down, and gaining speed. Pretty soon I started thinking, I sure wish we'd hit a tree or something to stop us. All I could visualize is a cliff or a giant boulder we might crash into. It seemed like an eternity, but suddenly we leveled off, and gradually came to a stop. We started breathing again, our hearts slowed down and we pried our hands from the grips we had taken. We gingerly got out, surveyed our location, and looked up the hill from where we'd come. Our tracks were straight as an arrow, missing huge trees by inches and squeezing between snags and boulders with no room to spare. The bench where we stopped was just wide enough to stop us before we would have started another long slide. We only slid about 300 feet, but it was the longest ride I'd ever taken, adding a couple of years to my age.

Tom rode with me lots of times afterward, but he no longer had unbounding faith in my driving abilities. Fact is, he got out and walked when things looked a little shakey.

Not all mishaps with Tom involved the snow cat. We left the ski bowl on skis one day in a white-out, heading for Gray Butte. We got just out of sight of the bowl in blowing snow, with no sense of depth perception, or even whether we were moving or not. Suddenly Tom dropped off a ledge about twelve feet high into deep powdery snow. Before I knew what happened, I skied off the same ledge, without even realizing I was moving. Tom had hurt his leg in the fall so we headed back for the ski bowl. If Tom weren't so tough, he'd be dead.

Rick, a young local man had notified the sheriff's office that there was a sick man in the North Fork snow survey cabin. He and Oliver had gone up to the cabin two days before on snowmobiles. One machine had run out of gas, and Oliver had come down with a toothache about the same time. They decided Oliver would stay at the cabin while Rick brought the other snowmobile in to town for gas. In a hurry to get to town, Rick ran off the road about four miles from the cabin. He worked hours getting back on the road. Another few miles and again he went off the road. This time he felt it would be easier to go down hill across country until he hit the road again. He finally ran out of gas in the heavy loose snow, and still had several miles to go on foot. When he did reach the road, the snowmobile tracks wouldn't support his weight, and he floundered for many hours before reaching dry roadbed near Lake Siskiyou. He caught a ride to town and was so bushed that he went to bed until late afternoon. Unable to get another snowmobile, he called the sheriff to start a rescue mission for Oliver, who he was afraid might have started out on foot from the cabin.

It was 10 p.m. and snowing lightly when Ben, Bud the deputy, Rick and I left town with the snow cat. We had never taken the Thiokol into North Fork before, but had no trouble as we maneuvered the machine up the trail, across country and through dense thickets. Snow was deep enough that the creek was bridged over in most places, making crossing easy. Ben was driving, and doing a good job when there appeared in our bright headlights a knife-edge ridge of snow, with nothing but blackness ahead and on the sides. We stopped and looked the situation over. By straddling the ridge, I figured the snow would push away in front of the snow cat, and the tracks would

212 / Forests—Fires & Wild Things

have enough footing to go on across, provided we could keep it centered on the ridge. "Unh uh," says Ben, "I agree," says Bud, as I tried to coax Ben to drive across the narrow ridge of snow. "It really looks worse than it is," I told them. "The contrast between the white snow, and the blackness beyond just makes it look bad. I'll be right in front of you signaling which way to steer."—"I'll signal, you drive," says Ben as he climbed out of the tractor. Bud and Rick climbed out, too. I was right this time. It went across without a hitch and made a good road to come back on.

The rest of the trip was fabulous. The light snow fell softly through the trees ahead of us, reflecting brightly in the headlights. The hum of the engine, the gentle rocking of the machine and warmth in the cab was very soothing and restful, even at 2 a.m. when we reached the cabin. The snow had stopped and the stars overhead were the brightest I ever saw in the crisp, clear, air. A thin smoke was coming from the chimney. We stopped, got out and entered the warm cabin. Oliver was sleeping soundly in a warm sleeping bag. His toothache was gone, and he was equipped with food for several days. We had rescued a man who didn't know he needed rescuing. Rick was the one who had been in trouble, not Oliver.

This was probably the last rescue mission for Bud to actively participate in, as he was elected sheriff a few months later, and I'm sure as the new sheriff he would send a deputy on such missions. Maybe not, knowing Bud.

We started before daylight to rescue a sick man near Lake Helen. His son had left him there the afternoon before, when he became too sick to travel. By daylight we had the Thiokol well up the mountain toward our destination, but the snow had become so hard, the machine wouldn't hang onto the steep hill. We unloaded and started on foot for another thousand foot climb to where he was. Dick, the snow ranger, the man's son, two deputies and I were in the party, equipped with stretcher, toboggan, ropes and all the paraphenalia for a full-blown rescue. He, too, was O.K. Apparently his sickness was from altitude, and he was still in his downfilled sleeping bag when we arrived. Because of his weakened condition and the glare ice, he did need assistance to get down to the snow cat. It wasn't easy driving the machine downhill because of the icy surface, and at one place we slipped sideways for a couple of hundred feet. Since I was driving, I could see that we would stop sliding in a little saddle, so I wasn't particularly upset, but the passengers were in near panic when we stopped. They all breathed a sigh of relief, and the sick man commented that he thought he was going to die the afternoon before, and knew he was going to die this morning in the clutches of the rescue team.

It's not too bad to make a goof where no one can see you, but the morning the snow cat slipped off the truck, everyone was there to see it and make ridiculous comments.

I'd gone to work an hour early because the snow cat had to be loaded onto the truck before we left at eight o'clock. The temperature was about 10 degrees, and the ice on the bed of the stakeside was about an eighth inch thick, and hard as a rock. I backed the truck against the snowbank and drove the Thiokol onto the bed. The truck looked level, but I guess it wasn't quite so. The snow cat sat perfectly square on the truck when I shut off the engine and climbed out onto the tracks. While standing there, I suddenly became aware of the snow cat moving. Almost imperceptibly at first, then faster, it began sliding sideways on the icy bed of the truck. All I could do was stand there, ready to jump if it fell off. The left track went off, but the frame caught on the edge of the bed, with the cat balancing precariously on the truck. I very gingerly jumped to the ground and tried to get a jack under the frame before it slid any further. By this time, Lee, Bob M., Larry, Tom, Ben and all my fellow workers had arrived, grinning or laughing out loud at the predicament I was in. Advice, mostly bad, was given by all. Everyone thought it was bound to fall off unless we got a wrecker to assist us. That is, all but Tom thought this. "I'll drive it off for you," he said, and got right in and did just that. A level head like Tom's could see the true situation better than I could, because I was rattled from the experience. *Moral:* "Things ain't always as bad as they seem," but—"Sometimes they're worser."

Jim and Russ needed a ride to Gray Butte in the snow cat one day, and things went exceptionally well. We got through all the normal problem areas without a hitch. Then I guess I got careless right at the top. Now Gray Butte had kind of a rounded top, but gets steeper as you go down the slope. We were within twenty feet of the building, and as I tried to maneuver the rig, it kept sliding sideways down the hill. I didn't want to admit to them that we were in trouble, so I told them to go on and fix the radio while I got the snow cat straightened out. I finally got the machine leveled by digging tons of snow from beneath the up hill track, but then it was sitting in a hole that I couldn't drive out of without digging a ditch in the snow for the upper track. They finished working on the radio. "What's going on? Why all the digging?" they asked sarcastically. Well, we got out, but only by digging another 75 feet of ditch and having them hang out from the door like outriggers on a South Sea Island canoe. I used more care in my driving the rest of the trip.

A heavy snow caught us before we could close all the lookouts in 1963. Pearl had her car at Bradley, and she scurried down when the snow started, but it was a few days later before Tom and I headed up the road in the old M-7 snow tractor to close the shutters, bring down the radio and winterize the building. About 24 inches of powder snow had piled up by the time we got a half mile from our destination, and the snow cat was struggling—with the loose powder flying over the hood and around the engine. Tom was driving and I was in the back of the rig, wondering why he was having so much trouble keeping the machine in the road.

"Where are you going," I hollered, as the rig ran off the road and started tilting on its side. He didn't answer, but the machine did stop. He seemed to be in a daze—"What's wrong with you, Tom?" He didn't answer, but just sat there shaking his head. "Are you sick?" I asked, and he nodded his head. I helped him out of the vehicle. He stood in the deep snow and cold air taking in great gulps of fresh air. Soon he was O.K. "What happened, Tom, are you all right?" "Yes," he answered, "but I blacked out for a while." We contemplated the situation, and decided he got an overdose of carbon monoxide from the engine exhaust as the snow forced the gases into the cab. Within a few minutes he felt fine, and we backed the snow-cat onto the road where we continued to the lookout. Then the work began. The shutters were sagging under the weight of the snow, and horizontal "snowcicles" had built out into the wind for a foot or more on the south side of the building. All this had to be scraped off before closing the shutters, and placing the bucket over the chimney was not easy, with all the ice and snow buildup, in the brisk, forty-mile wind. A hot cup of coffee brewed while we were working took the chill from our bones before heading for home.

Any time the district ranger has gone on a snow survey, things have gone wrong. Maybe it's just coincidence, because things go wrong occasionally without him, but *always* with him. Now this wasn't a serious problem, but typical with the D.R. along. Tom J. and I, with Lee, the D.R., left early for Deadfall. Things were going well, though the snow fell heavier as we went up the road in the snow cat. We took Sweetwater and Parks Creek courses and were getting behind schedule as the tracks sank deeper and deeper into the new falling snow. We were still over a mile from Deadfall when the snow cat just couldn't get traction in the powder snow, which was billowing over the tracks and reducing visibility to near zero. Figuring that skiing would be faster, we set out in the knee deep powder. Lee wasn't in the best of shape and the going was tough, as we struggled toward the course. Well, we finally got there and got the measurements and headed downhill to the snow cat, reaching it about dark. Things were looking up as we pulled around the hill toward Parks Creek Divide, but suddenly we realized the headlights were pulling more amps than the generator was providing, and the lights were dimming rapidly. It wasn't too difficult staying in the deep tracks made on the way up, but the darkness didn't help the situation, as we proceeded without headlights. On reaching the summit, I decided to ski down the road ahead of the snow cat, because despite the darkness, the snow was perfect for skiing, and visibility wasn't too bad. My intentions were to only ski for a mile or two, but the snow was good, and my eyes became accustomed to the darkness so I kept going, leaving Tom and Lee and the snow cat far behind. About a mile from the truck I was gliding along on my skis, at about five miles an hour and saw two dark objects looming ahead of me in the snow. Thinking they were a couple of large rocks which had rolled off the cutbanks, I skied between them. As I did the "rocks" exploded into life. They were two deer standing side by side, and had neither seen or heard me coming. Their surprise was no greater than mine, as one leaped down the bank to my right, and the other leaped up the hill to my left, and my heart leaped into my mouth.

It was about 11 p.m. before we reached home. A long, long day, and the last snow survey Lee ever went on.

STILL MORE FIRES

Lightning flashed, and the thunder rolled throughout the Cascades in Washington. Little or no rain fell with the storms of 1970, and the forests were tinderboxes, awaiting the spark to ignite. Winter snows had been minimal, and even the heavy fuels were dry and light as balsa wood. Kim as fire boss, and I as line boss, left for Wenatchee, and the Safety Harbor Fire in late July, to take over command of the fire from totally bushed overhead. The fire had already burned for ten days when we arrived, and had blackened 17,000 acres of timber above Lake Chelan, a deep, long and narrow glacier-type lake squeezed betwen steeply rising mountains. The deep, blackish water made me think it could harbor the American counterpart of "Nessie," of Loch Ness fame.

A week or more before Kim and I arrived, a crew from Mt. Shasta and McCloud, plus the Northern California hot shots had been sent. On our first day I could hardly recognize the men from my own district. They were haggard, tired to the point of exhaustion, dirty, and far from being the well trained, organized crew I knew. Even the crew bosses, Tom and Dirk, looked like they had been dragged through the mill. They told stories about the extreme behavior of the fire since they arrived—being run out of areas almost daily; working twenty hours a day; sleeping on the firelines; all this, day after day. Believe me, these were tough, trained men who could fight fire, eat smoke and outlast most men while doing it, but they now looked like the losers of a military battle. They had done their job, though, and when their lines were tied in and held, they just collapsed as a unit to take a well-deserved rest. I made sure they didn't get an assignment the following day, and then arranged for them to be shipped home as soon as possible. If I'm not mistaken, that was the last year that Tom or Dirk went on fires in line jobs. They'd had enough.

It wasn't because Kim and I did anything special or spectacular that the fire was controlled two days later.—No, it was just that the weather changed for the better, and we carried out the plans made by our predecessors.

The nights became still and cold. So cold that hoselays froze, and it was impossible to keep warm, even in camp. Shelters sprung up made of parachutes, packing boxes and whatever else the men could scrounge, with large campfires everywhere. The fire overhead from the local area were good men, but unfortunately, didn't have the experience of California firemen in coping with the unusually dry and explosive conditions of the 1970 season in Washington.

Kim and I, along with other California firemen went home

after about ten days, but the season was just beginning for them, as fires continued to plague Washington until late fall. Charlie and the Northern California hot shots went back to the Entiat Fire a couple of weeks later, not too far from the Safety Harbor, where they worked equally hard and effectively. They gained national recognition for their fire work on these fires, and I know they deserved all the credit they received.

KLAMATH—OFF—WOOLEY CREEK

My friend Bob M. and I were on a lightning recon when Larry, the dispatcher, asked us to take a look on the Klamath since they were totally involved with the Off Fire. Lookouts all over the Klamath country were socked in with smoke and haze as we flew the Salmon River, Scott River and major drainages around Callahan, Etna and Fort Jones. We watched a crew out of Callahan find and attack a small, hot burning fire near Callahan, and then proceeded with our recon. Nothing new was found, but while we were flying, the Off began to put on an awesome display, as the inversion layer broke through with the increasing heat and lowering humidity. From several miles away, we gave the Off an appraisal, with the practiced eye of some sixty years experience between Bob and me. "Boy, I'm glad we're up here and not on that baby,—it don't look good from here." Bob agreed wholeheartedly with my ungram-matical statement.

We arrived back at Mott airstrip about the time Larry or Ralph called. "Bob, we want you to go to the Klamath as firing boss.". You know, they think I'm an expert at this sort of thing. By the time I drove to Somes Bar, it was night, like midnight. The infrared imagery pictures had just been received and read by the interpreter. As usual, I thought the interpreter's readings were as fouled up as the rest of the organization, but went along with him like everyone else did. After all, if a specialist is not sure of himself, he's not likely to admit it; just as I didn't want to show my ignorance as I looked at the photos, so I pretended I understood everything perfectly.

Lines were not yet prepared for firing, so I went out with Charlie and his Northern California hot shot crew to build a line from a road down a ridge into Wooley Creek. We worked hard until daylight down the steep slope. This was dense virgin forest, with a considerable amount of understory and vegetation. By daylight, the fire was perking up below us and making runs up the hill near our line. It soon became evident that the fire was spreading laterally faster than we could build line downhill, so we had to abandon the line before it was tied in at the bottom. Normally a fire in heavy timber would not burn and spread during the early daylight hours, but the preheating and predrying from the day before made a dangerous situation from which we had to retreat. A new line down another ridge was constructed during the day shift. It was fired out and held.

We rested a few hours in fire camp during the day, and then coptered into a spike camp on Wooley Creek in late afternoon. Camp was in a flat meadow on the banks of the creek in a spectacularly beautiful setting. Tall Douglas fir, steep forested canyon walls and the sparkling clear, swirling waters of Wooley Creek presented the scene of a pioneer setting. At the edge of the meadow was a weather beaten pole corral, and a beautifully built, large log cabin was tucked under the trees at the downstream end of the flat. Except for the helicopter, the only access to the camp was by trail, some twenty miles upstream, and seven miles downstream from camp. It was in this secluded spot that President Hoover came to spend a few days to fish and relax during his ordeal as president.

The tranquil setting became a beehive of activity as firefighters, inmates, Hot Shot crews, Forest Service overhead and even a packstring of mules descended on it. Noise from pumps, copters and people soon drowned the sounds of breezes, water and little creatures. Smells of cooking and smoke from campfires blended with the pall from the Off, which was creeping downhill toward camp, still a mile or more away.

Dave R. and I, still working with Charlie's hot shots were to fire out the north side of Wooley Creek, beginning at dark. It was good to be working with these men. The Northern California Hot Shots have built a well deserved reputation throughtout the entire West under Charlie's supervision. I have worked with them often over the years, and familiarity breeds respect when you see and work with men in tough fire situations. They're tops, and Dave, who was in a trainee position with me on this fire was as qualified as I to do the job. We hiked downstream a couple of miles to where we could see the fire creeping downhill toward us, then another quarter mile to where the fire was already down to the trail. Everything was in our favor; a still night, cool, with ideal burning conditions. Fuses were adequate firing devices, as we started back toward camp, with fire strung out behind us. Most of the time, the fires burned readily without too much heat, but as the backfires climbed the hill into the drier vegetation above the influence of the stream, a roar like a freight train could be heard. The slope on the opposite side of the canyon would light up like day from the glow. All went well, and by two a.m. we were nearing camp, where lines, hoselays, and pumps had been put in readiness to protect the cabin and camp. We completed the job about daylight with a sense of satisfaction.

During the day, because I was one of the senior officers I got involved in a little discipline problem when the inmate kitchen crews started straying to where the lady firemen were bathing in the creek and making ungentlemanly remarks to some of those in the chow line, especially to those following the men's example of leaving their shirt fronts unbuttoned. "Button up, or quit complaining," I had to tell a couple of bosomy babes. A few of the inmates were sent back to the main camp at the request of their guards.

This co-educational fire fighting seems to have affected me more than the co-eds. Late the next afternoon, Dave and I were checking downstream from camp to decide where we'd best stay to watch for spot fires during the night. The day shifts were coming off the line. In one crew of about twenty firefighters there were six women. It was easier to walk in the stream than along the bank, so many of the men had taken off their clothes and were holding them out of the water as they passed. Both men and women on the crew seemed oblivious to anything unusual. Dave and I looked at each other with our mouths wide open.

It was an easy night.

I guess you've become a "wheel" in fire control when they send you to the Angeles in a six passenger, 240 miles per hour, twin-engine executive-type airplane. All by yourself, that is. Then to be met at the airport by a taxi to take you to fire camp, really caps the trip. Sure, it built my ego, but I'm certain the fire would have had much the same end results without me. As a member of regional fire teams, this happened to me a couple of times—once as line boss and once as safety chief. Let me tell you about these "wheel type" jobs; in my own case, I feel I developed into a good fire strategist over the years. Not outstanding, but not bad, either. Reputations, good or bad, depend on luck. No matter how good a fire boss, line boss, or any line worker is, if the winds aren't right, and the burning conditions are critical, he's going to look bad. I've had more than my share of good luck as a line boss. When I arrived, the winds died, the humidity climbed, and the organization really got into high gear. Believe me, it wasn't my strategic skills or organization efforts that stopped the fires. It was the weather. I'm not being modest, just factual. My luck wasn't always good, but I've had my share of it. Of course, sometimes good luck is a result of hard work and taking advantage of good situations.

Baldy Village reminded me of an Alpine village in Europe, or one in the Colorado Rockies. Rustic homes and stores tucked between granite boulders, steep rocky slopes towering above the narrow twisting streets and, could you believe it on the Angeles, *big trees.* The six thousand foot elevations contributed to vegetation unlike most of Southern California. Since the fire had started here, "Village" was selected as its name; and it had taken off with an impressive fury, according to the townspeople.

Little or no control was evident on the fire on my arrival as safety chief. Another fire was threatening the communities of the front country, and when the Village started, there were no immediate resources available to fight it. The local fire department had done a good job in protecting buildings and homes within the village, and had succeeded in building a few chains of handline in a critical area behind the town.

With only a handful of Forest Service personnel working the fireline, I had time to make some personal observations regarding the real possibility of the fire making a run at the settlement, necessitating an evacuation. Also, I had time to check out Bear Canyon, where about twenty summer homes were in real danger of being burned out. The fire was already backing downhill into the canyon, and with the Santa Ana's and local erratic wind patterns, the probability was great that the canyon would go. Now this canyon was different from the rest of the community—there was no road to the twenty or so homes. Only a foot trail reached them, and I would estimate their values up to $75,000 each. There were owners and friends with backpacks, wheelbarrows, or just carrying valuables down the trail when I first walked in. Old men, women and children were loaded down with their most prized possession; maybe a picture, a vase, a sewing machine or some stuffed animals. At the same time leaving what appeared to me to be valuable antiques.

Firefighters began to arrive in increasing numbers and a Shasta-Trinity fire crew was assigned to build a handline through the timber above the houses and a hoselay was laid up the canyon to protect them. Strong winds, especially at night still made the fire a threat to the canyon. I had made a test run on foot to see how long it would take for people to evacuate the canyon if things went to pot, and talked to everyone up the canyon, asking their cooperation if an evacuation became necessary. I felt a real kindred concern for them. Had I owned one of those homes, I'd have been reluctant to leave, too, feeling that I'd stay to the last to protect my interest. After assuring them that an all out effort would be made to save the houses in the event of trouble, they all agreed to evacuate on a pre-arranged order by the Forest Service.

Weather gradually got better, and the danger passed as the firelines held the fire. It was Thanksgiving morning. A slow drizzle started from a gray overcast, and by noon there was snow falling at Baldy Village, and heavy rains in the lower reaches of the fire. Everyone was joyous with the weather change, even though there were still a few miles of uncompleted fireline, mostly covered by snow. With an assist from the weather, the fire was controlled on Thanksgiving Day.

Fire camp was a sight to behold. Tents were hastily thrown up, cardboard boxes used for shelters, plastic sheeting everywhere and a "high rise" apartment of five foot culverts, piled like logs, sheltered dozens of fire fighters. The Thanksgiving dinner was terrific. Turkey, dressing, cranberry sauce, vegetables, coffee and all the other trimmings. Hot apple pies, made by housewives in the surrounding communities topped the elegant meal. There were groups standing in the chow lines singing in the falling rain, "Come Ye Thankful People," and "We Gather Together to Ask the Lord's Blessings."

The flight home was one I'll always remember too. Flying above the clouds wasn't bad. We could see thunderstorms going on below us all the way up the state. It was night, and we landed in several cities letting Forest Service people off. Going down through the clouds, fog and rain wasn't my favorite pastime, especially at Eureka where the fog reached the ground. Even the stewardesses seemed nervous. The Redding hot shots and I got off in Redding.

This was the year of two Thanksgiving dinners for me, as my wife had a full blown dinner for me on reaching home. No standing in the rain for this one.

Air drops occasionally go astray. Sometimes because of a mixup in instructions, sometimes due to pilot error, sometimes because of erratic winds and sometimes on the wrong fire.

A drop on a fire in the Sacramento River Canyon resulted in one soggy blonde in a red convertible. She, along with dozens of other cars on Interstate 5 had been stopped in a traffic control situation with air tankers dropping on a railroad fire adjacent to the highway. It was a hot dry day, and I'm sure she had looked pretty "classy" in her red Oldsmobile as she pulled to a stop—but not for long. For some reason, the bulk of the misguided drop hit right down the middle of the highway, mainly on "Blondie's Convertible." She looked like she'd been dipped in Pepto-Bismol, and the convertible was a mess.

"Who's running this show?" she asked Aaron, the transport driver. He pointed to me, "See that guy with the radio, he's in charge." She was irate, or worse. "Who's going to pay for cleaning my car and luggage?" I stammered, and told her to send a bill to the U.S. Forest Service in Mt. Shasta. She wasn't satisfied with my advice, but about that time the stalled traffic was turned loose by the California Highway Patrol, so she climbed back into the convertible, and drove away, looking very unglamorous. We never got a bill from her, but I'm sure she never forgot the experience.

Another time, the airborne tanker was not needed, but had to drop his load before landing, so the pilot decided to test his skill by using the Dome of Castle Crags as his target. He made a perfect drop, and the granite dome was "shocking pink" for the rest of the summer.

COLUMBUS DAY—1962

The Columbus Day, 1962 storm was one to remember. Southerly winds up to 75 miles per hour whipped through the town of Mt. Shasta. Trees bowed their tops almost parallel with the ground. Shingles peeled from the roofs, and rain fell in bucketsful. Power poles snapped as trees fell across lines and across housetops. For several hours the winds continued, with 100 plus miles per hour recorded at the Mt. Shasta Ski Bowl. Deer hunters were caught on back roads with trees criss-crossing the roads ahead and behind them. Some said trees fell right in front of their cars as they were driving. The full impact of the storm on the forest was not realized for several days, as foresters fought their way through hundreds of downed trees to appraise damages. My main concern was for the lookout structures, as fortunately, an earlier rain about three days before had brought the women off the peaks. At the first break in the storm, I'd looked with binoculars at Black Butte. I could see snow and ice through openings in the windswept clouds and fog, but the structure looked strangely different from a vantage point on the Everitt Highway on Mt. Shasta. I finally decided the peaked roof was missing. Hoping for a real break in the storm, I waited two days before going up to appraise the damages. Well, it didn't get much better, so I decided to go on up anyway. Some of Cecilia's belongings were still up there, plus a Forest Service radio, binoculars and other items. Don and Rich were with me, and by the time we reached the half-way point, the trail was slippery with ice and frost, and an icy wind was whipping sleet past us. We were tough—We were foolish—We continued. The last quarter mile was pretty bad. Six inches of snow, plus all the other miserable conditions probably made the chill factor about -10 degrees, and if there's such a thing as a comfort factor, I'm sure it was off the bottom of the scale. We continued—I turned around once to see if Don and Rich were still with me. They were on their hands and knees, braced against the wind, and their stocking caps were crusted with ice and snow, as were their eyebrows. We finally reached the lookout. It was worse than expected. The entire roof was gone, three fourths of the catwalk was gone, one wall was blown in, and broken glass was everywhere. Snow covered the firefinder, and a 14 foot timber which had been part of the ceiling structure and also served as an outrigger support for the shutters, was lying across it. The shutters which were still up when the storm came, were blown away with the roof. Beneath the floor, which was still intact, there was a dry basement where we went to escape the wind and cold, and to rest before packing our knapsacks for the trip down. We also dragged in a tank of bottled gas which we cracked the valve on and lit with a match. A hot pot of coffee brewed over the roaring gas jet soon had us warm and in good spirits. I do know the Forest Service safety handbook frowns on such as this, but we did it and were glad of it.

The trip down was easier, even with the heavy packs, but the real work was just to begin. Next week began the job of tearing down what the storm failed to destroy. On five consecutive, beautiful fall days, four of us hiked up the hill to clean up and burn up the wreckage from the storm. It's surprising what practice will do for a man. During the five days, our hiking time to the lookout dropped from an hour and fifteen minutes the first day to 45 minutes the last day, and no one really objected to the two and a half mile climb.

Pictures I took of the frosty trail and the wrecked lookout turned out well.

During the same week, Forest Service and industry foresters were appraising damage to timber stands throughout Northern California. If nothing else good came from the storm, it really put bureaucracy into high gear—line running, salvage sales and cooperative road building ventures were worked up in record times between the Forest Service, International Paper, Southern Pacific, U.S. Plywood and a host of other lumber companies. Existing sales were put to bed in order to get the blowdown salvaged before insects, rot and fire took its toll. Sometimes "Crisis Management" ain't bad.

HIPPIE DAYS

A few years ago, a staff reporter, and special writer of the *Los Angeles Times* and other newspapers throughout the United States interviewed me prior to writing a story on the hippie culture related to Mt. Shasta and surrounding localities. I was reluctant to be interviewed, but did finally agree to after some pressure. What came out in print in the *Times, The Atlanta Constitution*, a Milwaukee paper and no telling where else, was not what I had said nor what I meant. It was basically true but with totally untrue statements throughout the article. I felt embarassed by parts of the article supposedly quoting me, and to me it seemed offensive to certain local organized groups. Despite this, I feel the group generally refered to as "Hippie" makes a worthwhile contribution to this book.

They were mostly misguided young people, frantically looking for a different or even better standard of existing. Some were trying to get away from cities, the law, their parents' domination or indifference to them, or even to find a religious experience. Maybe to find the freedom proclaimed in these United States.

No, I didn't like them,—their ways of living, their moral standards, their scraggly hair or the aroma of their unwashed bodies. I'm sure my displeasure showed sometimes in my dealings with them as a forest officer. At the same time, I felt sorrow, sympathy, compassion and somewhere deep within me a responsibility toward helping them. Some were defiant, some pathetic; others silently pleading for a kind word or smile from someone.

One frail young man of about 19 built a shanty on Forest Service land near the Methodist Camp on Scott Camp Creek. He intended to live off the land by eating dried apples, wild berries and whatever he could find around the area. When we found his dwelling and gave him a deadline for tearing it down and restoring the area to its original condition, his mental anguish was very real and very disturbing to me. I'm not sure he ever understood why he couldn't "homestead" the area.

A young woman, also about 19, was camped at McGinnis Spring with her baby girl of about 18 months. The camp was dusty, and everything in it was filthy, including the little blond girl who was sitting in the lean-to, as bare as the day she was born, putting dirt in her mouth and hair. Mud streaks had dried on her cheeks where tears had run through the dust. I hadn't said anything to the mother, who was equally dirty, but she volunteered to me in a defensive voice that dirt would not hurt her child. No attempt had been made to cover up garbage or human waste around the camp. I made her clean up the area and told her I was going to report her to the county health and welfare department because of the baby. This I did, and within a few hours a county officer visited the camp, but I never heard what was done about the baby.

On occasion an obviously undernourished and spaced-out young girl would appear at the receptionist's desk of the Forest Service, and stand there glassy-eyed and unsteady on her feet until we called the local police to take her to the hospital. I believe her friends had brought her there and left her for the Forest Service to take care of.

Some hippies survived a cold snowy winter on Mt. Shasta a few years back. By building a small but sturdy shelter of slabs, bark, boards, limbs and tin, plus the addition of a heavy snowpack, two people spent the winter, but it was late summer before we found the structure, which we dismantled and burned. They must have been very resourceful; more so than the run-of-the-mill type.

Of course, too, there was the pair who camped one winter for a while in the toilet at Panther Meadow campground, evidently using the men's side as living quarters and the ladies' side for which it was designed.

Classical music doesn't always turn me on, but on a late summer day my wife and I were seeking a little quietness and solitude near Panther Meadow. As we walked the Gray Butte trail we began to hear violin music, faint but clear through the thin mountain air. Soon we saw a young woman standing on a huge boulder in the late afternoon sun. The music from her violin made chills up and down our backs, it was so beautiful. Undoubtedly she was a trained musician, and we wondered what had brought such a talented person to the point of living or existing as she did in the hippie environment then prevalent in the area. She obviously had brought some of her old culture with her in the form of classical music. Betty and I listened, and then walked on without being seen. When we returned, she had disappeared into the woods, probably to one of the hippy campsites scattered around the meadows.

Sometimes, when the job pressures get heavy, and nothing is going right, just getting away from the office for a while is a great remedy. One day while doing this, I wandered into a hippy campsite. The campfire was smoking and a pot of "something" was warming over the fire. "Here's a chance to get my thoughts together while waiting for the owner of the unattended campfire to arrive," I mumbled, as I settled down on top of a rock about 30 feet from the camp. The first arrival was a lean, hungry, nondescript looking dog of dubious pedigree. He meandered over to the pot of "something" and dipped his nose into it. A few bites were enough for him before continuing his aimless journey. Next, a lean, hungry, nondescript looking young man with shaggy beard and furtive glances stopped at the campsite. He evidently didn't own the site, but nevertheless, helped himself to some of the stuff in the pot, diping in with his fingers. After a few bites, he too, left, but returned in a few minutes to eat some more. "Well, anything is better than nothing," seemed to be his philosophy. Neither he nor the dog ever saw me, though I was sitting in plain view. It was another 30 minutes before the owner arrived. By this time the campfire had gone out. He saw me immediately. "What are you doing?" he asked. I told him of the still burning campfire when I arrived, and warned him to be more careful in the future. He did have a campfire permit, and had made an unsuccessful attempt to extinguish it before leaving some two hours earlier, so I did not issue a citation. The experience watching the dog and the man had bouyed my feelings, and I did not tell the owner about either as he started to finish off the pot. I was happy that he didn't ask me to dinner.

FOREST MANAGEMENT—WITH FIRE

Contrary to the thoughts of the public, Smokey Bear doesn't feel that all fire in the forest is bad fire. Within the past twenty years, especially since the early 1960's, there has been more acreage burned intentionally on the Shasta-Trinity Forest than by all the wildfires during this period.

Controlled burning, slash burning, crush and burn, windrow and burn, broadcast burning, clear cut and burn, pile and burn; it's all been done, and is still going on with varying results. Everything from unbelievable success to total disaster has happened. Reasons for using fire in forest managemnet differ with locations and circumstances. Our projects on the Sacramento (Mt. Shasta) District beginning in 1964 were to convert the dense brushfields on the southerly slopes of Mt. Shasta into productive timberlands by use of fire. Unsuccessful planting attempts had been made in the 30's by clearing narrow strips through the manzanita in order to plant trees, but the remaining brush soon crowded out the young pines to the point that they either died or became stunted as the brush competed with them for the moisture and soil nutrients. Our thoughts were to remove all the brush by crushing and broadcast burning, or by piling the brush into long windrows for burning before the next planting season.

Our first project was a 46 acre crushed brush plot just off the Everitt Memorial Highway, which went well for us, despite the fact that our guide lines for burning were based more on seat-of-the-pants judgment than scientific evaluations. Sure, we knew what the winds, humidity, and fuel moisture conditions were, but didn't know exactly what tolerances we had, or how much fire we could cope with. Up until this time, no definite guidelines had been established by the experiment station within which we eventually conducted our burns. The principles of crushing brush prior to burning it are based on creating a situation where the treated area will burn well when the surrounding area won't, thus making a plantable area without endangering the surrounding forest. The crushed and broken vegetation dries rapidly, and within a few weeks it reaches an explosive condition, where it burns with an all-consuming intensity, even in comparatively low fire danger situations. Weather conditions are watched, fuel conditions checked, and elaborate plans are made and approved prior to burning. With a constant weather pattern or trend evident, the date is set, and crews are alerted for a certain date and time of day to be at the burn area. Tankers, dozers, firing personnel, and holding crews are placed at pre-planned points to await the signal to start firing. A forecast for the day and the few days following is studied, and the current weather is checked to be sure the guideline limits for wind velocity, humidity, spread index and ignition index are reached before the firing begins. Usually the firing is done as rapidly as possible with drip torches, fusees, propane flame throwers or other devices. Rapid firing creates intense heat to totally consume the treated vegetation and also to utilize the ideal conditions for burning, since most large jobs are done in late afternoon or evening. Sometimes weather conditions change so rapidly that burning will almost cease with humidity increases; or conversely, if unpredicted winds arise, you soon have more fire than you can handle, so it is important to get the job done while the conditions are right.

Besides the benefit of converting brushfields to productive forests, firefighters learn more about fire behavior from these controlled burns than they ever could in a fire behavior class, getting a grandstand view of the effect of wind, slope and fuel types on fire. Perimeter firing, stripfiring and area ignition techniques have been and are used with varying success throughout Northern California for broadcast burning.

There is always a calculated risk in using fire as a tool in forestry, and I've had two serious mishaps, both in 1967. It was a bad year for me. Maybe the success I'd enjoyed for the several years before had made me a little less cautious, or maybe my luck just ran out. Anyway, I'll tell you first about the "Oops."

It was a beautiful mid-May morning. Conditions were borderline, in that we were crowding our burning guidelines all the way. Even so, I decided that with all our expertise and plans, the worst that could happen would be a few spot fires which we could pick up without too much trouble. After all, the snow had not been off the adjacent brush very long, and there were still a few snowbanks on the shady side of the trees on the uphill side of the burn.

Underestimating the potential intensity and rate of spread, I unwisely started firing the wrong end of the area, figuring we would need the extra heat to insure a clean burn. Now, Tom J. is not a very scientific fire behavior specialist, but he does know that fires burn uphill faster than downhill, and that a tail wind is going to spread it faster, still. "Bob," he said, "do you know that this fire is going right across the line and up that draw if we light it the way you want us to?"—"Aw, Tom, you're a pessimist; we can hold it with the tankers and Don's dozer."

Tom was right on all counts. Within an hour we were in bad trouble. The fire didn't just spot across the line. It roared across, and headed up the draw with the fury of a tornado. I was right about one thing. It burned clean, both inside and outside the planned area. We managed to hold the north end at the Everitt Memorial Highway by picking up a few spots across the highway, but not before burning about twenty acres of private land, and several acres of Forest Service plantation adjacent to the area. It also burned a few scattered trees in a brush patch not within the planned burn. Overall, a total of nearly 300 acres was consumed by the escaped fire, which finally stopped at the snowbanks in the timber above the brush.

I could really not blame anyone but myself for the escaped fire. Lee, the ranger who named the fire, was very understanding, and graciously accepted part of the responsibility. Tom still has never said, "I told you so."

The "Oops" has haunted me these past ten years. Fortunately, in most instances my friends and neighbors needle me in friendly fashion. "Remember the Oops!" they gleefully remark as we talk about controlled burning and fires in general. Some good came of it; we learned and established more secure guidelines for burning—and the deer hunting was never better.

One of the more spectacular, and successful burns was the Bear Springs Unit, several hundred acres of dense crushed brush surrounded by tall timber. It was dark by the time firing began, with ideal burning conditions.

Rapid firing along and around the perimeter soon created a spectacular fire. Smoke billowed high and straight up for several thousand feet. False cumulus clouds formed over the smoke column, and more than one person reported lightning in the cloud. Nearer the ground surface, an even more spectacular display was taking place. Whirlwinds and fire storms roared through the burning block, meandering from side to side, with flames shooting 500 feet into the air, occasionally crossing our control lines before dying down or moving back into the burn. One of our tanker crews had a near miss as a whirlwind headed toward them and crossed the line about 50 feet behind the scrambling tanker. This particular whirlwind deposited burning wood, brush and other fiery materials into the top branches of tall trees adjacent to the line, which eventually burned or died out before dropping to the ground. Needless to say, we checked the area thoroughly the next day for smoldering fires in the litter under the trees.

I was particularly happy that this had been a successful burn, as Or, the local newspaper editor had predicted in his paper that, based on our past two burns, there was a two-to-one chance that his property and cabin nearby would be burned when our fire escaped. He had some justification for his thoughts—though in reality the timber and fuel types surrounding our project were of a much less flammable nature than where we had two escapes the year before.

Tourists and sightseers have the knack of getting themselves in predicaments on occasion, and about three days after the burn, a young lady came to my office. She was well dressed, but dirty, and her face had tear-streaks through the dirt. "Mr. Gray," she said, "my Volkswagen is trapped up where you had the controlled burn; I drove down the hill to look at the area and can't get out." I knew exactly what happened.

There was a steep road downhill into the burn, with good roads within the burn. She had driven down O.K., and had driven the good roads in the V.W., but the loose dirt and rocks had been too much for her to drive out, so she had left her car and walked back to town for help. I was the help, so I drove her back out in my Forest Service pickup to appraise the situation. Sure enough, she was trapped, but we had built a very rough, indistinct road from the burn to the Wagon Camp road which I showed her, and told her to try with the V.W., while I waited at the junction. About thirty minutes later I could hear the V.W. roaring up the hill through the brush (if V.W.'s can roar), and soon she appeared at the top. She stopped her car and got out, dirtier than before, and immediately broke into gushing tears, of joy, I think. "Oh, thank you," she sobbed, "I don't know how to pay you for this, because I don't have any money, and my husband is going to be awfully mad when he hears about it." As I told her not to worry, and that I was just glad she was out, she threw her arms around my neck and planted a dirty, muddy kiss to my lips. "That's all I have to offer," she said, as I backpedaled to regain my composure. "That's enough reward," I assured her, and got her headed toward town in the dust-covered "beetle."

"The Indians let it burn." "The Indians didn't put fires out." "The Indians torched off the forest every fall before leaving for the low country." "It looked like a park when the Indians lived in it."—All these statements are quoted by advocates of the "fire is a natural part of growing trees" philosophy. And, you know what? They aren't entirely wrong. Neither are they entirely right.

Sure, they let 'em burn—They didn't use lumber, because they didn't have saws and sawmills. No, they didn't put fires out either; after all, fires created better browse for more deer; consequently, better hunting. And too, they knew that the fires would go out when they reached the ridge top or the river. Maybe they did set fire to the forests in the fall when winter was approaching. Could it be that's how some of the impenetrable brushfields recorded in survey notes of the mid-1860's were caused? If the forests looked like a park when the Indians were running things, they probably would have starved to death. Did you ever see any game animals in a parklike area of the forests?

There is also much written and said about the success in burning out the underbrush in the pine forests of the Deep South. That's where I'm from, you know. Let me tell you about those southern loblollys and slash pines. They're tougher than anything in the way of trees that grow in the West, and they grow where the rain comes down every month of the year. If you look at the whorls of branches, you'll find them up to three feet apart for each year's growth, compared to the six to twelve inch growth on their Western cousins, thus making it more difficult for the ground fire to climb the ladder of limbs to the tops of the trees. The situation in the West is just much more difficult than in other parts of the country.

I'm sure I'm not the first to have nearly insurmountable problems in trying to burn litter in timbered areas without damage to the reproduction and overstory, but I feel that I'm being honest and factual in saying it can't be done safely or economically. Let me cite what has happened when I've tried using fire to clean up litter in forested areas.

First, burning conditions which are perfect for doing the job, last for maybe two hours on a given fall or spring day, before becoming too critical to burn, or becoming unsuited for burning at all. Except for an occasional calm, overcast day, conditions vary to extremes from daylight 'til dusk, with temperatures fluctuating more than thirty degrees; humidity varying by 40 to 50 percent; and winds rising, falling and changing direction throughout the day.

In October, 1975, Ben, Robert, Eddie, Jack, I and a few others undertook a "clean-by-fire" operation of about 25 acres in the Sanford Pass area. If ever there was an ideal situation, this was it. An inch of rain had fallen a week earlier, and the day we chose to burn was calm and overcast. The fuels just right— dry enough to burn but damp enough not to overheat the timber and reproduction. A dry snow started falling lightly during the afternoon, but had little effect on the burning. We fired from a road at the top of the ridge, because the fire burned too hot where we tried uphill burning. Needles, duff and litter were about six inches deep, with a few small logs scattered through the area. It took several days to complete the job because at times in mid-day the fire intensified to the point we would have to slow the downhill spread to prevent sporadic crowning in pole-size pine and Douglas fir. Finally, on about the sixth day, we put control lines on one side of the burn, as strong winds and hot afternoons combined to create a serious threat to the adjacent timber stand. All the time, burning was done within a predetermined prescription, and when the prescription components were exceeded, all burning was stopped.

The results? Not good, so far as I was concerned. Materials on the ground were not consumed by the fire; just a charred mess was left. Too many trees were killed; there were areas where the convected heat from light litter completely desicated the needles on 30-foot trees. Heat penetrated the heavy bark on the lower trunk of larger trees to the point that the cambium layer was dried on two thirds of the diameter. I feel that insects and disease will kill these trees within two or three years as a result of the creeping fire. And the cost? About $60 per acre, with as few men as possible, in ideal conditions for burning. We took many calculated risks by cutting personnel to a bare minimum in order to lessen the cost.

In another area, completely surrounded by cleared plantations, John W., a forester on our district, and I burned the litter beneath about a quarter acre of mixed pine and fir under slow burning conditions with the same results. Many of the standing trees crowned out and died from the fire, without really accomplishing our purpose. If we're going to clean up the woods and create a parklike setting in cutover or standing timber, it has to be some other way than by broadcast burning.

SAINT GERMAIN FOUNDATION

According to the Saint Germain Foundation, or the "I Am'ers" as known to most people in our area, Mt. Shasta was where St. Germain appeared with a message to Mr. Ballard, who started the religious order bearing the name of Saint Germain. Whether one believes as the "I Am'ers" believe is a personal choice, but I have enjoyed my friends who are members of the St. Germain organization. I first met Mrs. Ballard at the "I Am" headquarters in the old Shasta Springs Resort area between Dunsmuir and Mt. Shasta some years after her husband's death. She impressed me with her aura of authority as I explained why we were driving fire trucks through St. Germain property. A fire on the railroad was most easily accessible by using the foot trail from the headquarters to the fire near the Shasta Mineral Springs some three hundred feet below on the railroad. She was very cordial, and told me we could come through the property any time for offical business. She had a bone crunching grip as I shook hands upon leaving. On our return from the fire several hours later, cold soda pop was served the thirsty firemen by members of the group.

I have found "I Am'ers," and I don't use the term as a derogatory reference, to be of a kind and gentle nature; rather intelligent; and good workers. Some have been employed by the Forest Service in office jobs, as lookouts, and a few in field jobs. Katherine, who has served many seasons as a lookout on the Shasta-Trinity, is among my older friends, a highly respected lady.

One of their members, an older gentleman, maybe 80 years old, provided me with a humorous few moments. It was a hot, dry summer afternoon, when I was driving along a road on the outskirts of town that I noticed a man burning a stump in a dry grassy field. I knew no one had issued a burning permit, especially under the critical conditions at the time, so I walked over to where he was sitting, watching his fire. "I'm sure you don't have a burning permit, do you?" I asked, as I introduced myself. "Why no, I didn't know I needed one, could you give me one?"

The next few minutes revealed that he was just burning the stump out for a friend who owned the land. "No, the friend hadn't asked me to, I just thought it would be a nice thing to do for him," he told me. "Well, we'll just have to put this out; don't you know that you could get into serious trouble if a wind came up and made the fire get away?" After a pause, he said, "See that mountain?" as he pointed to Mt. Shasta. "All I have to do is look over there if I get in trouble, and help will come." I was getting a little exasperated as I said sarcastically, "And, I suppose help always comes, huh?" "You're here, aren't you?" were his final words. We put the fire out, and he promised me no more stump burning on hot summer afternoons.

I seriously doubt that all members of the St. Germain Society believe exactly as my friend, though he seemed sure of the reason for my appearing when I did.

SAFETYWISE & OTHERWISE

Dear Bob:

After checking our accident reports and safety goals for the Shasta Forest, we find that in the year 1946 just ending, you have had three lost time accidents. First, a cut and infected thumb from working telephone lines; then a wrist injury from a chainsaw, and last, an injury requiring medical treatment after getting your hand caught in a gopher trap. This frequency is about 10 times the acceptable level for forest workers.

How do you account for such a high injury frequency?

Yours very truly,
Al
Safety Officer
Shasta National Forest

Such was the gist of a letter I received in the early days of my Forest Service career. I'm not sure they wanted an answer in return, but here's how I responded to it.

Dear Al:
    The only explanation I can come up with is that I do about ten times as much work as the average forest worker.

> Yours very truly,
> Bob

Seriously, though, the work of the Forest Service is hazardous, and it is remarkable to me that more and more serious injuires don't occur. The types of work, and working conditions are so varied, it is nearly impossible to set up a fail-safe safety program. Safety slogans, safety plans, safety posters and signs; no matter what you use, results are about the same. Safety meetings, tailgate safety sessions, or no safety at all—it seems to matter not at all what kind of a year it'll be safetywise. And too, within the last ten to fifteen years, people made accident reports on injuries that no one would think twice about in the 40's or 50's. A minor cut would take care of itself; poison oak would cure itself about as quickly without going to a doctor; a week of limping would take care of a twisted ankle; why, most anything less than a broken bone or a severe cut would cure itself with time, and that's the way lots of it was done way back in the 40's.

With a background of the aforementioned injuries to me, plus many others not mentioned, I became a safety chief on regional fire teams in the mid 70's. Bob M., my good friend said, "You ought to be a good one, you've violated most of the safety rules for thirty years, so you sure know what to look for." You know, I think he was right. Except for the first fire I acted as safety chief on, we had fantastic success from the safety standpoint, and despite Bob's implications, I think I did a good job. Let me tell you the first experience—it really shook my confidence.

It was the Virgin Fire on the Trinity side of our forest. I thought maybe a reminder list posted beside the chowline, showing all the possible ways to get hurt, might serve as a warning of things to watch for as they worked on the fireline. My strategy backfired. Of the thirty-three ways to get hurt, as shown on the list, they were all accomplished within 48 hours. Plus a few others not on the list. Everyone must have thought it was a goal to reach, rather than things to avoid. No one was killed, but the first-aid line and the ambulance trips to the Weaverville Hospital continued day and night. My own people from Mt. Shasta were the worst offenders, with four of ten being sent home with injuries.

Four of my personal friends have been killed on fires; Andy and Forrest in the Inaja Fire disaster in 1958, Chris on a lightning fire on the McCloud Ranger District, and Sam, in a helicopter accident on the Sequoia. Others, who I didn't know quite as well, were killed in other copter accidents, from rolling a bulldozer while unloading from a transport and one from bee stings. I reserve a special place in my heart and mind for these men.

On fires, the most dreaded accident is being trapped, unable to escape an oncoming blaze. To this end, fire training is stressed—in classrooms and on the fireline, with special emphasis on extreme fire behavior due to weather, topography and fuels. Situations indicating potential danger are spelled out in the "Situations Shouting Watch Out" transparencies and slides in both Basic and Intermediate Fire Behavior training. My own lone experience in getting trapped was on the Mears Fire of the Modoc, over thirty years ago. Fuels were fairly light, but extreme weather had caused blowup conditions after 10 a.m. almost daily. I had taken a couple of men in a Jeep to corral a spot fire about two hundred yards outside the fireline.

When we decided we couldn't hold the spot, we headed back to the safety of the fireline, but in the meantime, the fire had jumped the lines and a wall of flame was racing toward us, cutting our avenue of retreat. We whirled the Jeep around and headed to a large corral we had seen near the spot fire, which we reached in time to sit out the oncoming flames. There was nothing to burn in the corral except the fence, but the heat, smoke and dust made the twenty-minute wait seem like an eternity. I never told the sector boss of my experience.

Ralph, the Shasta Forest F.C.O., and Fred, the district ranger at Fall River Mills had a rather hair raising experience on the Porcupine Fire in 1950, when they drove a Forest Service sedan into an area which later became blocked off at both ends by fire. Their plight was known to all, because they had radio communications to the rest of the fire, and it was several hours before we or they knew what their fate was to be. Fortunately, they both came out with nothing worse than a little less hair on their heads. The evidence still shows—Just kidding guys.

For every accident where someone gets hurt, there are probably dozens of near misses. Roger G. told me about this one. Seems he and a few men were working on a small fire at either Porcupine or Picayune Lake in the early forties. Roger was skirting around the edge of the lake with a portable Pacific marine pump on a packboard on his back, searching for a suitable spot to set it up for use on the fire. Suddenly his feet slipped, and he fell down the rocks into deep, deep water. Now those early day Pacific pumps weren't light, and Roger and pump immediately sank to the bottom of the cold, clear lake. Roger laughed about the incident when telling it to me some thirty years after it happened, but it wasn't even slightly humorous to him at the time. He didn't tell me how they retrieved the pump after he managed to get out of the packboard.

I happened to be with John S. when a five hundred pound flaming treetop plopped to the ground within inches of him on Gray Butte, and saw an almost identical incident happen to his brother James a few years later. Both were without warning and were on lightning fires. I'm sure both John and James remember the incidents even more vividly than I.

It was a hot, windy afternoon when Francile spotted a railroad fire near Graham. It was beginning to heat up pretty well before I reached it, so we decided to request an air tanker from Montague. I was only about four miles away when she reported it, and I reached the scene before anyone else. That F7F must have already been warmed up, because it arrived only a minute or two after I did. It made a low approach run over the length of the narrow fire before circling for its actual drop. Since not being very effective with my shovel, I got out of the way to let him drop. It was a good thing too, because he zoomed in at high speed, and low elevation—about thirty feet above the grass, sagebrush and light manzanita. His load hit that fire in one big "lump" of retardant. It uprooted everything, and rolled it into a ball of vegetation at the head of the fire, and I'd have been wrapped up in that ball had I chosen to continue working where I was. It was the most effective air drop I've ever seen, putting all the fire out, and mopping up most of the smokes.

No, we don't tolerate that kind of air tanker use, but I guess a hot airplane, and a hot shot pilot just couldn't resist the temptation to show off a little bit. I later talked to the pilot who agreed he was flying a little low, and would refrain from doing so in the future.

THE BIGFOOT INCIDENT

It happened this way: Walt, the deputy sheriff was in my office discussing a criminal case with me, when a telephone call came in. It seems that the wife of Virgil, a faller on the Whirlbird timber sale, received a call from him on her C.B. radio, asking that the sheriff's office be notified to send a deputy to the sale area. Virgil had seen a seven foot "beast" walk by him in the woods, then stop and look at him from a distance of less than 50 feet. The sighting was accompanied by a terrible odor, like an old rotten bear hide. "Be sure to bring a

Bigfoot, as described by logger.

gun, there is dried blood on the brush near the place I saw him," he added to the message.

"I'll go with you, Walt;" I grabbed my hard hat. Now Walt had his street shoes on, so had to go home first to change into his hiking boots. After an awfully long wait he got back to the office, and we took off up the road toward the sale area. I'm pretty impatient about such things as dawdling when there's a crisis afoot, but not Walt. Everything was in apple pie order by the time we were ready to talk to Virgil and the woods boss.

Virgil told us this story: "I had left my saw and equipment on the hill the night before where I finished work. It was about a forty minute walk into the area and I arrived at the saw about 8 o'clock. Being a little tired from the hike, I sat down to rest a few minutes before going to work. Then I heard a sound of something coming down the hill toward me. It was a shadowy, upright figure, and I thought it was either Rex, the ranger, or another faller walking toward me. I called 'over here,' to let him know where I was. No answer, but I could still see it coming downhill. When it got about the same level as I was on the hill, and about 30 feet away, I called again. This time the creature stopped, turned and looked straight at me over some brush. It had to be seven feet tall, and had hair all over him from the top of the head down to below his shoulders, which was all I could see. The hair was not shaggy, but swept back over his head, and the face and eyes were dark below the hair. It looked at me for a few seconds in good daylight, then turned away as it continued walking down the hill. There was a terrible odor which hung in the air after it walked away. I then went to find the other faller to tell him what I'd seen. He came back to the scene with me. We could still smell the thing, and then we found some dried blood on some brush, like something had been killed and dragged away. By this time we were pretty edgy, so we cut a club from a fir sapling and headed back to the landing."

Aside from Virgil, the woods boss told Walt that if any one but Virgil had told this story to him, he'd have just thought, well, he's been drinking again. "But not Virgil. He's a level-headed, dependable, honest man, and I know he saw something," he said.

After listening to this story, Virgil, Walt, Rex and I took off up the hill on foot to the scene. Walt was not in the best of shape, so it took us an hour or so to reach the site, where Virgil pointed out the location of the creature, the stump he was sitting on when he saw him, and then showed us the blood on the brush.

We reenacted the scene, with me sitting where Virgil had been, and Rex going over to where the creature stopped and looked at him. In order for me to see where the face of the creature had been, Rex had to hold his hard hat up to a height of over seven feet from the ground. Some of the men thought they could still smell faint traces of the odor described by Virgil. We found footprints in the needles and litter where the creature had walked, but nothing very distinct. The "blood" turned out to be a red fungus stain on the brush, but did look like blood. A sample of some droppings found in the area still remain a mystery. The lab was unable to identify it, but were sure it wasn't from a bear, coyote, man or other creature common to the locality.

I'm sure Virgil saw something, and as he again described the figure to Walt, I listened from a few feet away, while sketching what I heard him describe.

"Is this what it looked like?" I asked, handing him the sketch.

"Yes," he replied. "Really, he looked a lot like the characters on the 'Planet of the Apes' T.V. program."

Rumors ran rampant for several weeks. "It was a forest ranger with a Smokey Bear mask," went one. Another, "It was another logger playing a trick on his co-workers," or, "It was hippies trying to scare the loggers away because they didn't want the area logged." The newspapers purposely, or accidently quoted Virgil as saying, "The stench was horrible, I first thought it was a forest ranger." So far, the mystery is unsolved; or maybe there is a Sasquatch on Mt. Shasta.

I was two hours late for lunch that day, and after explaining why to Betty she replied, "Well, I've heard some good excuses these 28 years, but this is probably the most unique."

TOWARD THE END

It was in or about my twenty-eighth year with the Forest Service that I began thinking of retirement. Not because I felt physically unable to do the work, nor was it that my mental facilities had deteriorated too far. It was just that the Forest Service was changing its ways, policies, ideas and processes, while I resisted the trend. Of course, in my own mind we, the Forest Service, were doing some ridiculous things. For instance—lady firemen. I was appalled. "When I have women on my fire crews, I'm through," I declared, but lived to eat my words, as I had at least seven or eight the next few years. There was Cheryl, Jana, Terri, Nan, Sue, Lynn and one or two other short-timers. Grudgingly, I'll admit some did an acceptable job, but as a whole they were more of a liability than an asset. I didn't dislike them, just didn't want them on my fire crews. Despite their awareness of my feelings, I hope they choose to remain my friends.

Then there was the Forest Service pre-attack plan. Try as I sincerely did in its initial stages, I could not justify its costs, effort, time required or anything else about it—except for its intent, which was the same as the fire control intent before its inception. Southern California, where fires cover large areas each year, is more suited for the pre-attack concept than the Shasta-Trinity, and I have effectively used parts of the pre-attack set-up on fires in the Angeles Forest.

How about the sophisticated recruiting system in use by the service the last three years. Applications were computerized, the information fed into automatic data processing machines, and *presto!* The names, grades and priorities came out on a long list, in order to be hired. I'm afraid it picked the best educated and the ones with the most talent for filling out applications. What we really needed were the ones who have average intelligence plus the ability to pull hose, throw dirt, chop brush, fall trees, be tough and durable, and put out fires. It hurt my ego to think the "machine" could do a better job of recruiting than I, a mere human. Don't know how we survived the years when we recruited by choosing those applicants who had to be either a good ball player, or to duck his head when he entered the door for an interview.

What really hastened my decision that the service and I should sever our relations was the management by objectives program. Now M.B.O. was not new to me—I'd practiced it throughout my career, and very well, too, I might add. The only difference was that they wanted it on paper, and not in *my* head. Not on just a piece of paper either; they wanted it on reams of paper, plus flow charts, plus progress charts, plus safety plans and alternate plans. To the point that the plans were bigger than the projects.

And last but not least—how to manage all that money. Everyone worried about my finances but me. I knew how much money I had, how much I needed, and how and when to spend it. That wasn't enough. "Your figures don't agree with the printout, Bob," says Val or Bonnie. "Neither do the printout figures agree with mine, so I figure the printout needs correcting," was my truthful answer. Well, at the end of the fiscal year I was always O.K., despite my quarrels with the computer and the office girls, bless their hearts. When I told them their worry over finances would cause gray hairs, they chorused, "It's not the finances, it's *you* that causes gray hairs."

CONCLUSION

Yes, there are lots of people, characters, friends and acquaintances mentioned in my adventures or misadventures. But there are many more that I worked with and for, who just happened to not be involved in incidents related in this book. They also helped make my "ranger career" the most satisfying job imaginable. Each in his or her own way helped me develop character, understanding, patience, knowledge and all things necessary for a good life.

My district ranger bosses were many, and of divers personality and philosophy. Dutch was one of a kind—as indicated in the early chapters of this book. Bob T., and Bob M. were young, ambitious and competent. I only worked with them for short periods, but respected their ability and decisions, even in later years as our paths crossed on fires. Fred was the fatherly type; very easy to work for and very efficient in his running of the district. Even at his retirement party he was referred to affectionately as "The Godfather."

Horace—an energetic, nervous, active type ranger, who would as likely be out doing a yeoman job of trail work with the crew as attending to the managerial chores in his office. Those could be and were done after hours. I sometimes thought my easier-going attitude bothered Horace. I only hope it helped calm him somewhat.

Frank and Dave were acting rangers, a decided disadvantage to them, as their employees tended to disregard their position or authority. They have remained my friends, and have served the Forest Service well. I hope I helped them in the short time I worked for them.

It was with some apprehension that I began work for John, who turned out to be an excellent boss; a pleasure to work for. Any reluctance I had to work for him soon vanished as we developed a friendship and understanding. Jim had been John's assistant ranger, so there was no trouble getting used to him. Little change was made in district operations, but I was never sure Jim liked being ranger.

Lee had grown up on the Sacramento District on fire crews and tankers. It was like the kid brother coming back—only now I was working for him, as were Bob M. and others on the district. His understanding of fire control and its related problems, plus his acceptance of any idiosyncrasies I had, made my job easier.

Big changes came in the Forest Service soon after Ron arrived. A new forest supervisor, new policy changes, new programs of work, and a "new ranger." Either I failed to understand the new ways, or subconsciously didn't want to understand them. Ron and I had a few rough years, with clashes over everything—personnel, civil rights, job planning, management procedures, fund accountability. You name it, we had problems in understanding each other's philosophy, methods and ideas. Ron had more of a commitment for the changes than I could muster, but, despite my lack of enthusiasm for some of the programs, I tried to accomplish the desired *results*, which I felt were more important than the *methods*. Ron and I have had a better relationship since my retirement than before, in fact, a mutual friendship.

Lookouts have always been among my favorite Forest Service workers. Maybe because I started in that job. There was, in addition to those in other stories—Ralph Oiler, Paul McGinty, co-workers on the McCloud District in 1942. Barbara was on Little Mt. Hoffman in 1945; just out of high school, and a good lookout. My wife and I accidently ran into her and her husband in Susanville in 1977. A fine lady and gentleman.

Laverne was on Black Butte when her radio took a blast of lightning. She and Dick still live in Mt. Shasta and are among my favorite Forest Service friends. Wanda, also at Black Butte—a salty character who proclaimed via Forest Service radio that "Lookouts are not serviced; they're supplied—only cows are serviced." This when someone said he was on his way to service a lookout. Tough gal, that Wanda.

Then there was Roxie, a cute young lady, about whom Bob M. said, "I can see you didn't let Betty pick this one." One afternoon she called by radio to tell me her dog had been bitten on the nose by a rattlesnake, so I hurried up to Sims to get the dog to take to the vet. Pal sat very obediently on the seat next to me on the way down. All the time his nose was swelling bigger and bigger, until at last his head looked like a blimp. Sad and dejected he looked when we arrived at Dr. Tucker's. With two days rest and good treatment he was back to normal with no ill effects.

June recited poetry from Slate Mt. during the late visitor's hours, and later married the above mentioned Dr. Tucker.

"Aunt Toosie," believe it or not, was afraid of heights, like the lookout tower at Bear Mt. and Black Fox. After weeks of creeping up and down stairs on hands and knees she finally conquered her fear, or at least hid it. Became a good lookout too.

Seems like there was a new lookout each year at Black Butte. In order were Ann, Mildred and Cecilia, who didn't come back after the Columbus Day storm, but continued on Grizzly for the many years since. Then Roxie, Beth and Ione, with Walt stuck in there somewhere, rounded out the lookouts on Black Butte prior to it being removed to Hogback Mt. on the Shasta Lake District in 1974.

Marge and Jerry manned Sims during a season when Lenore traveled Europe, and enjoyed Lenore's summer home immensely.

Peg became a mainstay on Bradley. Working in the ranger office in winter and at the lookout in summer kept her well versed in Forest Service business.

Del at Weaver Bally was one of the best—also one of the most independent—persons I ever met. We got along fine, except one morning she was called by Al, from fire weather, to give some weather observations. Sure it was before working hours, but! "Who's going to pay me for this extra work?" she wanted to know. I hustled right up to the lookout to inform her that she wasn't even going to get paid for any more regular work with an attitude like that. "Why, Del, I average at least two extra hours a day—for free in my work, and I don't think it'll hurt you to answer a few questions before work." We didn't really have harsh words, but they were effective (until I left Weaverville).

It was natural to become acquainted with neighboring district lookouts, too. Cala, at Herd Peak, Dorothy on Plummer Peak, Betty on Bonanza King, Barbara on Girard, and several on Shasta Lake District—lookouts became known by voice and location. Especially Rosy on Sugarloaf. A terrific person and lookout, who I am sure never sleeps. Any hour of the night, her pleasant voice is talking to someone on the radio. Either about a shooting in the campground; a extra large campfire on an island in the lake; directing me or someone to a smoke along the railroad track; or contacting someone for the dispatchers in Redding to send to Southern California. "I talk in my sleep," she assured me.

Even Jim and Joan, my brother-in-law and sister, spent their honeymoon on Limedyke, and another season on Hayfork Bally, while both were in college. A couple of characters who did their jobs well, according to Art and Glenn, their bosses. Joan even put out a fire started by some soldiers working near Hayfork Bally. She was more than mildly upset at the U.S. Army.

Of forest fire management officers; I went through a few during my three plus decades. Having good relationships with all, and learning much from each one. I hope I was able to help them, though I'm sure some thought, that Bob just won't listen to reason sometimes. Maybe even, I was a stumbling block to some of them. You know—in pre-attack, fuels management philosophy, and like. All of them helped me in my career, and I'm grateful to them; Ray Huber, Ralph Bangsberg, Bob Flynn, Ed Heilman, Monroe Kimsey, Jack Godden, George Mendel and Ray Trygar; each an individual with lots to offer in different ways.

Busy people, forest supervisors are, with demands far exceeding available time, but every one took some time with me and my job. Not much with Mr. Davis, as I was on a lookout while he ran the Shasta. Norm Farrell was one of the guys, even borrowed my rifle to hunt, and came back to report that it didn't shoot straight. He'd missed a big buck just out of Bartle hunting with Dutch. "I told you it needed sighting in before you left here, didn't I?" "Yes," he admitted, but hadn't bothered to do so.

Bob Jones O.K.'d a couple of jobs to help me get established in the Forest Service for which I'm grateful. Then there was Paul Stathem. In my moment of crisis with the escaped controlled burn, better known as the Shasta Fire, he defended my position, which at best was rather awkward, before the Regional Board of Review of 1967 fires.

Dick Philf inherited, acquired and worked into a tremendous job as supervisor of the Shasta-Trinity National Forest, and most of my dealings with him have been indirect, but with good feelings, and I appreciate the difficult task he is doing.

There was Bill C., about 17 years of age, who brought me a hen pheasant (illegal) as a peace offering after doing a lousy job painting windows in my Forest Service residence in McCloud. He's now the assistant fire management officer on the Klamath, working with George Mc., and other Klamath firemen who are my friends in the Yreka area.

I especially remember Bob G., the McCloud district dispatcher, Bob and Alva N., dispatcher, crew foreman, and later with the experiment station—these, along with Mel D., Dick S., C. Sherman, Merv. P., Ollie H. and a host of other district dispatchers from all over the forest. Several of these men and women later went on to greater things with the Forest Service.

There were mechanics and shop foremen who I remember well. Ernie E. once told me, "We don't use tankers as bulldozers, Bob," as he straightened the front bumper. Harvey F. looked at my tie rod every time I drove up, since our first meeting was to straighten out the pretzel shaped tie-rod on my new pickup. J. D., Sam W., Roy B., Dick B., Itchy, Darrel, Glen and Bill, plus those at the Northern California service center are all remembered for their cooperation and help to me over the years.

There's no way to forget the staff personnel and clerical personnel on districts and in the supervisors office. George S., Andy S., Dana, Paul R., Ed W., Ora Mae, Ethel, Elsa, Francis, Flossie, Gracie—to name a few way back when, plus more recent ones like Val, Bonnie, Norma, Ione, Kathy, Pat, Rosie, Judy, Mary, Roberta, John M., Dave N., Jack K., Duane T., Don W., Bill L, Jim H., Jim P. and an almost unending list of foresters, technicians and clerks who came and went with the seasons. There are friends at the service center in Redding, any of whom worked with me. Don S. is now running the place, Bob H. is a wheel with the helicopter program—he had ulcers working with me and I told him I thought he should be causing them—not getting them. These, plus the jumpers, pilots, mechanics and warehousemen have a big place in my memories of the Forest Service.

People and events continually come to my mind from long ago and of recent years. My thanks to all who made an unforgetable event of my years as a "Forest Ranger."

In November of 1976, I reluctantly agreed to have a retirement party in my honor. Now, I've always enjoyed other people's parties, but thought, This is not for me.

How wrong I was. The throngs of people who filled the old city park recreation hall to wish me well, overwhelmed me. People I'd not seen in years showed up—from the F.S., C.D.F., private industry, and just friends and relations. I hardly remember what happened, but the good (and bad) stories, the friendly atmosphere, and the good fellowship, the letters from people unable to attend, and the fun enjoyed by all, made me realize how good a life I'd had for the past 32 years. These people were real friends whom I love and respect. This was one of the real highlights of a wonderful career. Lee, my ex-boss, was the M.C., and there was great participation in the fun making, from the speeches, through the dance 'til the wee hours. And nobody went home early.

Linda, Peg, and Becky did all the hard work, with help from all over. Their production was a masterpiece in getting things so well organized. My thanks to you, and all who came. It was a wonderful ending to a long career.

Back Cover:

Burning trees near Black Butte, between Weed and Mt. Shasta, California. September 1972.

This humorous and very informative account about the Forest Service from 1942 to 1976 was written by Bob Gray, a 33 year veteran of the Service. It contains nearly 200 true stories about events occurring on the Shasta-Trinity National Forest and forest fires throughout California, Oregon, and Washington State.

This book gives a close personal look at subjects the ranger deals with, from forest fires to forest administration. Wildlife, conservation, humorous incidents and unique characters are covered, along with a few tragic happenings, and many other well told true tales.